Origin and History of Marginal and Inland Seas

Proceedings of the
27th International Geological Congress

Proceedings of the
27th International Geological Congress
Moscow 4–14 August 1984

Volume 23

ORIGIN AND HISTORY OF MARGINAL AND INLAND SEAS

VNU SCIENCE PRESS
Utrecht, The Netherlands
1984

VNU Science Press BV
P.O. Box 2073
3500 GB Utrecht
The Netherlands

First published 1984

ISBN 90–6764–032–8 Volume 23
ISBN 90–6764–009–3 set of 23 volumes

Printed in Great Britain by J. W. Arrowsmith Ltd, Bristol

PREFACE

The leading scientists in the many different fields of geology were invited by the Organizing Committee to present a paper on a specific problem in present-day geological science at the 27th International Geological Congress. The published proceedings of the Congress consist of twenty-three volumes. Each volume is dedicated to a particular aspect of geology. Together the volumes contain all of the contributions presented at the Congress.

The Organizing Committee is pleased to acknowledge the efforts of all of the participating scientists in helping to produce these proceedings.

Professor N. A. BOGDANOV
General-Secretary of the
Organizing Committee

CONTENTS

GEOLOGY OF THE BERING SEA DEEP BASINS

BOGDANOV N.A. (1), NEPROCHNOV Yu.P. (2)
1. Institute of the Lithosphere, the USSR Academy of Sciences, Moscow, USSR. 2. Institute of Oceanology named after P.P.Shirsov, the USSR Academy of Sciences, Moscow, USSR.

Water face of the Bering Sea, the northmost of the Pacific margin seas, is located between North America and Asia. These two continents, at least during Cenozoic had not been separated by depression with oceanic lithosphere and now they form homogeneous entity. Within vast shelf in the northern part of the Bering Sea the structures of the two continents are not divided by meridianal lineaments and judging by magnetic field, can be traced from east to west, from America to Asia. The shelf of the Bering Sea occupies nearly one half of the total water face along 180° meridian and its width reaches 1000 km.

Deep basins of the Bering Sea - the Comandorian in the west and Aleutian in the east are located in the southern part. They are separated from the Pacific plate by the Aleutian island arc and the Aleutian trough. Along the boundary of the Aleutian deep basin and the Bering Sea shelf there is a steep continental escarp, cut by deep canyons, most of them are related to the zones of vertical disruptive disturbances. In the north and west the Comandorian deep basin is separated from Asia by continental shelf and steep continental escarp. Judging by magnetic field, this boundary has a character of vertical fault without noticeable horizontal dislocations, as linear anomalies from the deep basins are continued in the Olutor zone of Koriakia. The Aleutian and Comandorian deep basins are separated by the Shirshov-Bowers undersea ridge system.

Thus, geographic position of basins of the deep part of the Bering Sea are different, what, without any doubt, can be explained by the age difference and difference on

Proceedings of the 27th International
Geological Congress, Volume 23, pp. 1–17
ORIGIN AND HISTORY OF MARGINAL AND INLAND SEAS

geological evolution of these large blocks of oceanic crust. Let us consider the geological structure of the most important elements of the Southern part of the Bering Sea. While analyzing the material, apart from the published data, we have used the results of the recent geological and geophysical survey carried out during 29th leg of SRV "Dmitri Mendeleev" (neprochnov, 1983), fig.1 and the results of works being done by the scholars of the Institute of the Lithosphere of the USSR Academy of Sciences in Northern Kamchatka and Southern Koriakia.

THE ALEUTIAN DEEP BASIN

The Aleutian deep basin occupies the south-eastern part of the Bering Sea. As a result of detailed geophysical research carried out in this region during the last twenty years (Ewing M. et al., 1965; Stone, 1968; Scholl et al., 1975; Gainanov et al. 1968; Ludwig, 1974 etc.) oceanic crust with thicker sedimentary cover has been established. Under horizontal sediments of the basin the burried ridges of the basement are discovered. The most significant of them are the Bartlet and Sounders ridges, wihich are 2-2,5 km higher than surrounding areas. Thickness of the sedimentary cover in the central part of the basin reaches 3,5-5 km, toward the peripheries it increases up to 7-9 km. It achieves its maximum (about 11 km) at the foot of the Bowers ridge. Mohorovicic discontinuity under the basin is traced at the depth of 15-16 km, i.e. 3-4 km lower than under the Pacific plate, adjacent the Aleutian arc. The seismic section of the Aleutian basin's crust reveals bed with velocities 1,9;2,9;4,3;5,4;7,0 Km/sec.

Geomagnetic researches in the Aleutian basin helped to identify meridianal magnetic anomalies M1-M13, the age of which increases from east to west. These anomalies are oriented nearly perpendiculary to the Mesozoic magnetic anomalies of the Pacific plate, traced to the south from the Aleutian trough. To our opinion their nature is doubtful. Most probably they correspond to the ridges of acoustic basement, which have tectonic origin.

The age of acoustic basement of the Aleutian is not stated definitely. Judging by the interpretation of magnetic anomalies Cooper (Cooper et al., 1976) defined the age of formation of the basin's basement as 137-117 mln years or Early Cretaceous, bearing in mind that the oceanic plate, forming the foundation of the basin, had been, at that time, separated from the Pacific ocean by the Aleutian arc. The probability of such an approach, in their opinion is supported by data about the character of heat flow (1,42-1,34 UHF), and the type of gravitational field. They also have similar parameters with the northern part of the Pacific plate. We suppose that the process of divergence of the Aleutian and Pacific plates took place in the end of the Cretaceous, thus the age of the Aleutian basin's basement also can be younger - i.e. Middle Cretaceous.

THE COMANDORIAN DEEP BASIN

The Comandorian basin is located in the south-western part of the Bering Sea. It is separated from the Aleutian basin by the system of the Shirshov-Bowers undersea ridges and from the Asian continent - by a relatively narrow shelf. It is rather typical that if geological and magnetic anomalies on the shelf of the Aleutian basin, in Alaska, as well as in Koriakia are parallel to the northern boundary and cross magnetic anomalies of the bed at nearly right angle - in the Comandorian basin magnetic anomalies af the bed extend northwards into the shelf and farther they are corresponded by this or that montaineous structure of the Olutor zone (Rghevski, 1977). In the far west lines of nappes are parallel to them. The nappes can be traced along the fornt of ophiolite sheets and other deepwater formations. Envelope of tectonic sheets stretches from the Govena peninsula southwards across the Karaginski island reaching the Kamchatka cape.

Acoustic basement of the basin is overlaid by 1,5-2,0 km thick sedimentary cover. According to the NSP data, the rocks of the sedimentary cover are represented by the upper well stratified and the lower - acoustically transparent horizons with seismic velocities of 1,83; 2,60; 3,10

5,75 km/sec.(Rabinowitz, Cooper, 1977). Sometimes the tops of the burried mountains **raise above** the sedimentary cover; under the cover the mountains form a system of linear and slightly curved ridges (Fig.2). Their amplitude of raising is 0,5-1,5 km. Approaching the Shirshov ridge, the basement of the basin, in general rises. (Gnibidenko, 1980).

The character of anomalous magnetic field allows to relate mafnetic anomalies to not deeply lying basic magmatic rocks of the basement. The depth of M discontinuity under the basin reaches 9-10 km, i.e. the thickness of the crust here can be compared with the normal thickness of oceanic crust.

The heat flow in the Comandorian basin is approximately 2,5 times higher than that in the Aleutian one. Its absolute meaning achieve 2,64-5,03 UHF, the minimal ones are fixed in the north-west, progressively they increase to south-west. According to Sugrobov and Smirnov (1979), such a heat flow can be explained by high position of 1100° isotherm above the Comandorian basin, what, in its turn, can be explained by a young age of the acoustic basement. This fact is also supported by anisotropy in the mantle (Snor, Fornari, 1976) under the basin, which evidences to the Tertiary spreading.

Section of the sedimentary cover of the basin was revealed by site 191 (Creager, Scholl et al.; 1978) in its eastern part, which, by its morphostructural properties is more similar to the slopes of the Shirshov ridge, than to the central basin. In this point the thickness of the sedimentary cover achieves 900 m. The upper part of the section is made up of interbedded distallic turbidites and aleuritic diatomaceous silts and clays as well as of fine grained sandstones and volcanic ashes of the Upper Miocene-Golocene. Dense argillites and aleurolites of the Middle Miocene prevails in the lower part. They lay on the diffuse surface of tholeiitic basalts, the age of which, defined radiometrically, is 29,6 mln years (Middle Oligocene). Taking into account the morphology of the bed and the given geophysical data, one can suppose that the basalts of south-western and western parts of the Comandorian basin are younger. The Shirshov ridge serves the key structure for the understanding of the history of information of the Bering Sea basins.

THE SHIRSHOV UNDERSEA RIDGE

The Shirshov ridge separates, as mentioned above, the Aleutian and Comandorian basins. It begins at the Aleutian arc and as a continuous system of mountains, is traced till the Olutor peninsula of Koriakia. The Olutor ridge is its extension on the continent. Thus, up to the valley of the Ilpy river there is a 900 km long, 50-80 km wide mountain system. The ridge rises 100-200 m above the bed of the adjacent basins, and, generally has asymetrical structure - its eastern slope is less steep than western one. The ridge consists of the row of parallel shains, the height of which increases in the axial zone. The whole ridge, excluding steep escarps of the western slopes, where their amplitude reaches 700-800 m, is covered by sediments, which in the mountain valley achieve thickness of 1-3 km. The structure of the sedimentary cover is complicated by plicative deformations and rupture dislocations. The formation of the last evidently is caused by the intensive current movements, characteristic for the whole western part of the Bering Sea.

Deep seismic sounding, carried out in the Shirshov ridge for the first time during the 29th leg of "Dmitri Mendeleev" (Neprochnov, 1983) allowed to reveal in its crust the sheets with velocities 1,8; 3,2; 4,0-4,5;5,0 and 6,8 km/s. The Mohorovicic discontinuity is characterized by velocity of 8 km/sec. The total thickness of the crust is 18-20 km, i.e. nearly two times thicker than that of the crust to the west (Fig.3).

Geomagnetic observations carried out during the 29th leg showed that the western part of the ridge is characterized by multiple magnetic anomalies while the eastern one, at least up to the depth of 4-5 km, is formed by non-magnetic rocks. These observations are explained during the processing of data, obtained from dredging.

During the 29th leg 13 dredgings have been done over the

western slope and axial part of the ridge. On the western slope outcrops of intensively tectonized and mylonitized gabbroids, dolerites and banded amphibolites have been discovered. The age of basalts from the latest determinations carried out by means of K/Ar method in the laboratories of the Institute of Ore Deposits, Petrography, Mineralogy and Geochemistry of the USSR Academy of Sciences is equal to 36$\pm$3 mln years (Middle Oligocene). Judging by the character of sediments -siliceous argillites, grits and sandstones of Miocene - tectonic deformation and mylonitization connected with the cover formation in the eastern part of the ridge, took place earlier. It is evident that olivine and olivine-pyroxene-plagioclase basalts are the youngest magmatic rocks here. According to Scholl''s data (Scholl et al.,1975) their age is determined as the Middle Miocene (16,8 mln years).

In the central part of the Shirshov ridge predominating of basalts and siliceous rocks was established by dredging. The last are represented by strongly lithified formations of grey, grey-greenish and violet colour. They differ in purity of their composition. These are fine grained quartz-chalcedony, chalcedony and crystaballitic-chalcedony rocks (Bogdanov, Vishnevskaya, Sukhov, 1983). Complete skeletons of radiolaria combine no more than 25% of rocks and about the same amount is presented by organogeneous detrites. The balk is made up by fine-grained siliceous and clay-siliceous matter. These rocks had been accumulated under calm conditions, close to those of abyssal plains. Good preservation of fine skeletons of Diatoms and fragile delicate forms of radiolaria witness to that. Microfauna complexes, washed out from the rocks speak for their Late Cretaceous (Campanian-Maestrichtian) - Early Paleocene age. Comparison of composition of the Upper Cretaceous flints of the Shirshov ridge with the rocks of the same age to the north and west from the ridge proves that they had been accumulated in the calmest and deepest parts of the bed being at a considerable distance from the areas of drift of clastic and volcanogeneous material.

The upper layers of the central and eastern part of the Shirshov ridge are traced towards the Aleutian basin and, probably within it serves as the upper horizon of the acoustic basement.

Thus the Shirshov ridge presents a jointing zone, a zone of oceanic piling up and a zone of subduction of two blocks with oceanic crust - the Aleutian plate, formed during the pre-Tertiary time and the Comandorian one which continued to form till the Middle Miocene.

Jointing nature of the ridge is well expressed on its northern extension within the Olutor zone.

STRUCTURE OF THE FOLDED FRAMING OF THE BERING SEA BASINS

The structure of the folded framing of the Bering Sea basins is rather complicated. Nappe folded elements of the western Alaska and northern Koriakia framing the Aleutian basin in the north have alot of common structural features and all over they are oriented in parallel to the continental escarp of the Aleutian basin. We have already mentioned that according to the data of magnetic survey the main folded forms within the shallow shelf are traced continually between the continents. Within the northern Koriakia as well as on the territory of Alaska wide spread of tectonic nappes, forming accreationary prisms has been established. Such microcontinents as Vrangelia and the Golden ridge massif, shallowwaters Palaeozoic and Mesozoic formations of continental slopes, ophiolite complexes of the Late Paleaozoic and Berrissian age make up such prisms. Zones of serpentine melange, olistostromes and other complexes spread widely within accreationary prisms had been formed due to horizontal tectonic movements. Another particularity is ratherinteresting and characteristic for this vast territory - formation of the Alaskan-Koriakian accreationary prism which occurred due to the shift of blocks from the south-east to the north-west and completed by the Albian. In these regions, heat-loving fauna of the Tethys type having nothing in common with the fauna of the North Eurasia and America of these time periods can be found in the Upper Palaeozoic and Trassic formations.

In Albian, along the edge of the accreationary prism separating, a terrigenious flysch like wedge, separating the prism from the oceanic plate began forming. Wide strip of flysch series, no less than 5-8 km thick is traced in nearly latitudinal direction in the interfluve of the Ilpy and Pikaswayam rivers in Koiakia. Farther it submerges near the edge of contemporary continental escarp of the Aleutian basin. The flysch belt, similar in age and composition is located on the American continent, i.e. the southern part of the Alaska peninsula. No doubt, that tectonic shifts, nappe formation, different types of rift formation and volcanism within the Alaskan-Koriakian accreationary prism occurred in later epochs, but as a structural element with a transitional type crust this prism, in general, completed formation by the Albian and further on it was not connected genetically with the Pacific oceanic plate.

In contrast to the Alaskan-Koriakian region, the structures of the Olutor zone of southern Koriakia are oriented meridiannaly and are submerged under the waters of the Comandorian basin (Bogdanov et al., 1982) the eastmost of them, the Olutor ridge orogtaphically is replaced by the undersea Shirshov ridge. The Upper Cretaceous and Lower Cenozoic rocks occurring in typical oceanic sequence made up by tholeiites and the Albian-Turonian siliceous formations are spread within the Olutor ridge. As for the character of volcanic formations (Sukhov, 1983) the Olutor ridge is made up, in general, of tholeiitic and lime-alkaline basalts of oceanic uprising and has a number of common features with the character of volcanism in the Imperor ridge and the New Hebrides. No doubt, it appeared as a result of oceanic piling, the Upper Cretaceous formations, similar to the rocks of the Aleutian basin basement formed olistostromes along the eastern edge of the ridge (Bogdanov, Vishnevskaya et al., 1982).

The Oligocene and Lower Miocene young volcanic formations make up the central part of the Olutor zone. Here, in the lower part of section the tholeiites dominate, higher up they are replaced by lime-alkaline and contrast series. The Middle Miocene-Pliocene olivine plateaubasalts present the youngest magmatic formations of this region.

In the west of the Olutor zone another Middle Miocene

accreationary prisms or the zone of oceanic piling is located. It can be traced from the Kamchatka cape peninsula across the Karaginsky island up to the ridges of Vivensky and Govena. The Middle Cretaceous ophiolites are spread here, among them amphibolites, close to the Shirshov ridge amphilolites are present. Back arc complexes take part in formation of the nappe structures of the region. Judging by a number of indirect signs one can suppose that the whole of the Karaginsky-Govenian accreationary prism is shifted over the edge of the continental plate, the Pre-Cambrian socle of which goes out in the Taygonos peninsula and Central Kamchatka. Taking into account the seismic activity of the region and fold formations in the youngest sedimentary series, one can consider that the process of piling here continued in the Quaternary and, therefore, the axis of spreading in the Comandorian basin being located closer to Kamchatka, than to the Shirshov ridge.

GENERAL CONCLUSIONS

Thus we have dwelt upon the main features of geological and geophysical structure of the deep basin of the Bering Sea and its framing. Summing up these data we can come to the following conclusions (Fig.4).

1. The acoustic basement of the Aleutian basin is formed by the oceanic plate of the same name, the formation of which completed in the Albian (about 100 mln years ago). According to feological data the mid-oceanic ridge during this geological epoch stretched in parallel to comtemporary continental margin and the Upper Cretaceous terrigeneous wedge. In connection with the above, there appears a question of interpretation of meridianal magnetic anomalies, established by Cooper et al. (Cooper, Marlow, Scholl 1968) in the central part of the Aleutian plate. They were correlated with anomalies M1-M13 according to the Larson scale (117-132mln years). In general magnetic anomalies are parallel to the uprisings of the acoustic basement and the burried ridges of Bartlett and Sounder,as

well as to the Shirshov ridge. It is evident that they respond to the transform faults. This conclusion is supported by an indirect proof: coincidence of anomalies with the direction of deep canyons (for example: the Pribilov's canyon) crossing the shelf. No doubt, that the canyons formed along the faults with shift dislocations. Conditionally we can suppose that the basin's sedimentary cover, subdivided according to the seismoacoustic properties into three horizons. Correspondingly they can be referred to the following ages: the lower one (velocities 3,4-3,9 km/sec) - to the Late Cretaceous-Eocene; the middle one (velocity of 2,7 km/sec) - to the Oligocene- Early Miocene; the upper one (velocity of 1,6 km/sec) - to the Miocene-Quaternary period (ben-Avraham, Cooper, 1981). The Aleutian plate splitted from the Pacific one in the end of the Cretaceous along the large transform zone, being a part of system of the East-Pacific transform lineament (Mendosino, Murrey, Cleeperon et al). In Tertiary time, as a result of subduction the transform fault zone transformed into the system of the Aleutian arc-trough, thus completing the isolation of the Bering Sea from the ocean.

2. The Comandorian basin differed considerably from the Aleutian one and the adjacent oceanic regions by its geological and geophysical properties. The basin's basement has deeply dissected young relief, the basin preserves high heat flow, anisotropy of mantle rocks is typical for these territories. All these factors and close structural relations of the basin and structures of the Olutor zone in Koriakia witness for quite a different mechanism of basin's formation. The Comandorian basin had formed as a result of local spreading, which began in the end of the Cretaceoustime and lasted till Late Miocene. The basin is asymetric - its younger part is narrower and located at the foot of Kamchatka's continental slope. The axis of spreading is shifted to the west and oriented meridianally and in parallel to the Lower Cretaceous transform faults. Therefore it is logical to suppose that it started forming along thelargest structure of such type. The spreading was compensated by the subduction of oceanic crust under the continental one (in the west); the process

on its first stage was accompanied by the formation of island-arc Andian Type volcanism, replaced in Oligocene and Miocene by the obduction and formation of the Vyvenkan-Karaginsky accreationally prism. In the east, in the zone of the Shirshov ridge occurred the subduction of a young (Late Cretaceous and Early Tertiary) oceanic crust of the Comandorian basin under the Middle Cretaceous (Albian) oceanic crust of the Aleutian plate. The subduction was accompanied by the formation of numerous horizontal breaks, zone of nappe structures and oceanic piling. In the Late Tertiary time the central part of the basin was filled by lime-alkaline volcanic rocks and sedimentary clastic series. Active spreading in the Comandorian basin, probably, caused the cessation of subduction in the western part of the Aleutian trough and transformation of the Benioff zone inthis region into rightsided shift. Since the Late Miocene time island arc volcanism and siesmic activity are not characteristic for the Comandor's islands. At the same time, earthquakes occurred along the continental margins - over the Govena peninsula, the Karaginsky island and the peninsula of the Kamchatka Cape. The Comandorian basin, by the mechanism of its formation is similar to the Parese-Vela basin in the Philippine Sea (Karig, 1972, Geology of the Philippine Sea, 1980).

3. The undersea Shirshov ridge and its continental extension - the Olutor ridge present the remnant volcanic arc which formed above the subduction zone and piling of two plates with oceanic crust. The Shitshov ridge, in the east is made up of oceanic tholeiites of the Albian and in the west - of the Middle Oligocene age. The Late Cretaceous volcanism of the ridge witnesses for the fact that the subduction of the Comandorian plate under the Aleutian one occurred along the zone of transform rift. Probably, the Shirshov ridge in the Early Cretaceous time presented the northern extension of the Imperor ridge and the rift started forming during this very geological epoch at its foot. As a result of spreading in the Comandorian basin the ridge moved eastwards along the Comandorian-Aleutian shift.

4. The epoch of contrast tectonic movements in the Bering Sea region best of all was expressed since the Middle

Oligocene. At this time large superpositioned basins - the Khatyrskaya, Navarinskaya, Anadyrskaya and others were formed and started intensive accumulation of terrigeneous sediments, derived from the continents. These basins may serve as large reservoirs for oil and gas.

REFERENCES

1. Bogdanov N.A., Vishnevskaya V.S., Sukhov A.N., Fedorchuk A.V., Tchekhovitch V.D. 1981. Oceanic olistostromes of the western coast of the Aleutian basin (the Bering Sea) Geotectonika N5, pp. 74-81.
2. Bogdanov N.A., Tchekhovitch V.D., Sukhov A.N., Vishnevskaya V.S., 1982. Tectonics of the Olutor Zone. In: "Review of tectonics of the Koriakian Plateau". Nauka, Moscow, pp.189-217.
3. Bogdanov N.A., Vishnevskaya V.S., Sukhov A.N. 1983. Upper Cretaceous formations of the undersea Shirshov ridge (The Bering Sea). The USSR Academy of Sciences, Annual, p.273, N 5.
4. Vishnevskaya B.S., Sukhov A.N., Tchekhovitch V.D. 1982. The age of the Vatyn series (the Olutor zone of the Koriakian Plateau) Review of the USSR Academy of Sciences, Geology N 12, pp.71-78.
5. Gainanov A.G., Kosminskaya I.P, Stroev P.A., 1968. Geophysical research of the Bering Sea deep structure. Review of the USSR Academy of Sciences, Sec. Physics of the Earth, N 8, pp.3-11.
6. Geology of the Philippine Sea bed. (under A.V.Peive), Moscow, Nauka, 1980, 261 p.
7. Gnibidenko G.S., 1980 Tectonics of the Comandorian Basin. In: Geology of the Comandorian basin, Vladivostok, pp.5-12.
8. Neprochnov Yu.P. 1983. 29 th leg of SRV "Dmitri Mendeleev", Oceanology, Vol.XXIII, is. 2, pp. 365-368.
9. Rghevsky N.N., Zatsepin E.N., Ustinov N.V., Shinarev V.N. 1977. Particularities of geological structure of the south-west part of the Bering Sea as for airomagnetic data. In: Problems of geophysical research

of the polar areas of the East. L., pp.14-25.
10. Sugrobov V.M., Smirnov Ja.B. 1974. Heat flow in the north-west of the Pacific. Nature N8.
11. Sukhov A.N. 1983. Volcanogeneous complex of the Olutor ridge. Review of the USSR Academy of Sciences, Geology ser., N 10.
12. Ben-Avrahamz, Cooper A.K. 1981. Early evolution of the Bering Sea by collision of oceanic rises and North Pacific subduction zones. Geol.Soc.Am.Bull. part I, v.92 pp. 485-495.
13. Cooper A.K., Marlow M.S., Scholl D.W. 1976. Mesozoic magnetic lineations in the Bering Sea marginal basin, Journ. Geophys. Res., v.81, No II, pp.1916-1934.
14. Creager J., Scholl D.W. et al. 1973. Initial Reports of the Deep Sea Drilling Project, Vol.19, US Gov. Printing Office, Washington D.C., 913 p.
15. Ewing M., Ludwig W.J., Ewing J. 1965. Oceanic structural history of the Bering Sea. Journ. Geophys. Res. v.70, pp.4593-4600.
16. Karig D.E. 1972. Remnant arcs. Geol.Soc.Amer. Bull. v.83, pp. 1057-1068.
17. Ludwig W.J. 1974. Structure of the Bering Sea basins. In: Burk C.A., Drake C.L. (eds.) "The geology of continental margins", New York, Springer, pp.661-668.
18. Ludwig W.J., Murauchi S., Den N., Ewing M., Hotta H., Houtz R., Yoshii T., Asanuma T., Hagiwara K., Sato T., Ando S. 1971. Structure of the Bowers ridge, Bering Sea. Journ. Geophys. Res. v.76, pp. 6350-6366.
19. Rabinowitz P., Cooper A.K. 1977. Structure and sediments distribution in the western Bering Sea. Marine Geology, v.24, pp.309-320.
20. Scholl D.W., Buffington E.C., Marlow M.S. 1975. Plate tectonics and the structural evolution of the Aleutian - Bering Sea region. In: The geophysic and geology of the Bering Sea region: Geol.Soc.Am. sp. paper. 151 pp. 1-31.
21. Shor A.A.Jr., Fornari D.J. 1976. Seismic refraction measurements in the Kamchatka basin, Western Bering Sea. Journ. Geophys. res. v.3, pp.5260-5266.
22. Stone D.B. 1977. Plate tectonics, palaeomagnetism and the tectonic history of the N.E. Pacific. Geophys. Surveys, v.3, pp.3-37.

SCALE 1:5000000

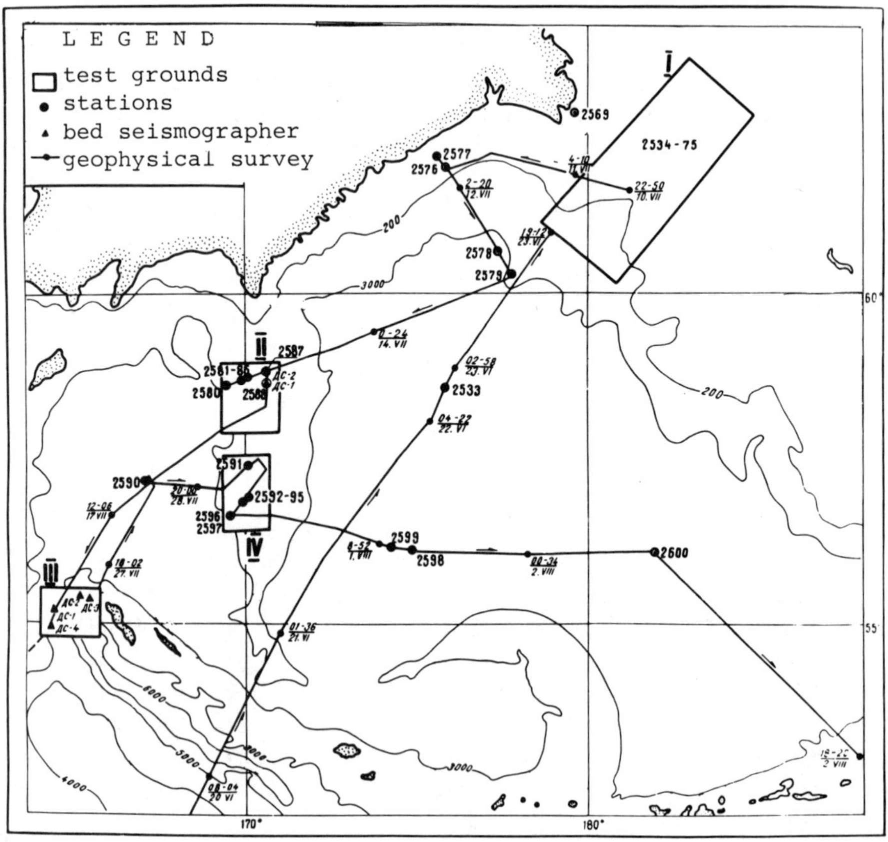

FIGURE 1. Scheme of the works of the 29th leg of "Dmitri Mendeleev" (1982).

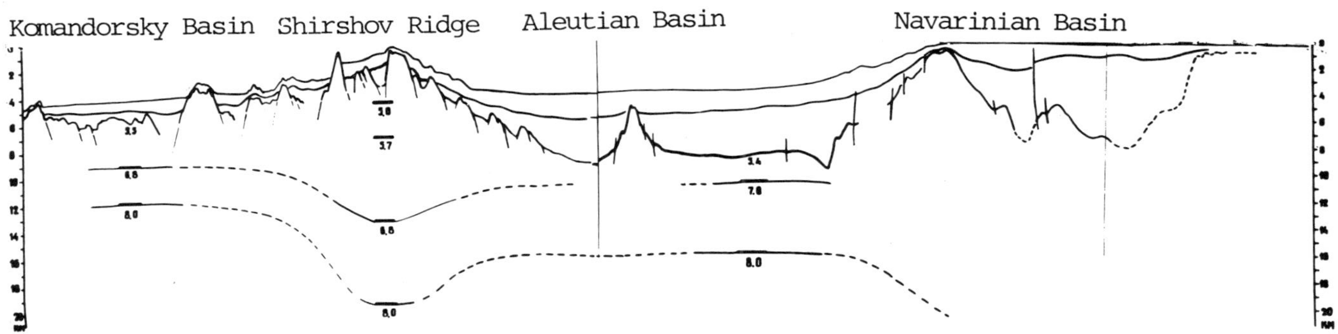

FIGURE 2. Section NSN across the Shirshov Ridge and adjacent part of the Comandorian basin.

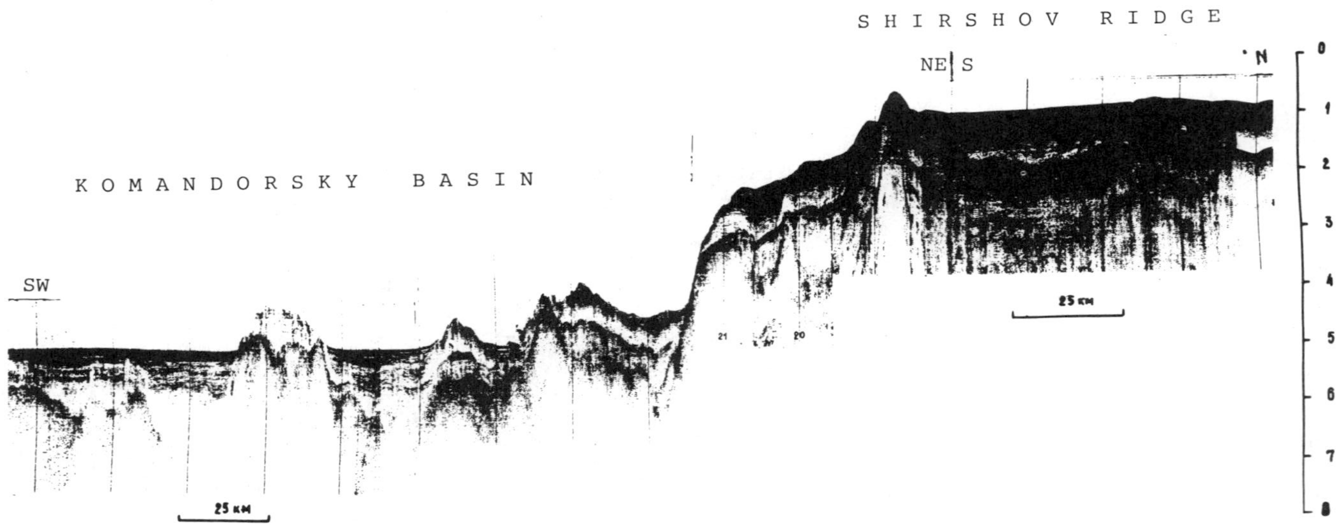

FIGURE 3. Combined seismic section across the Bering Sea.

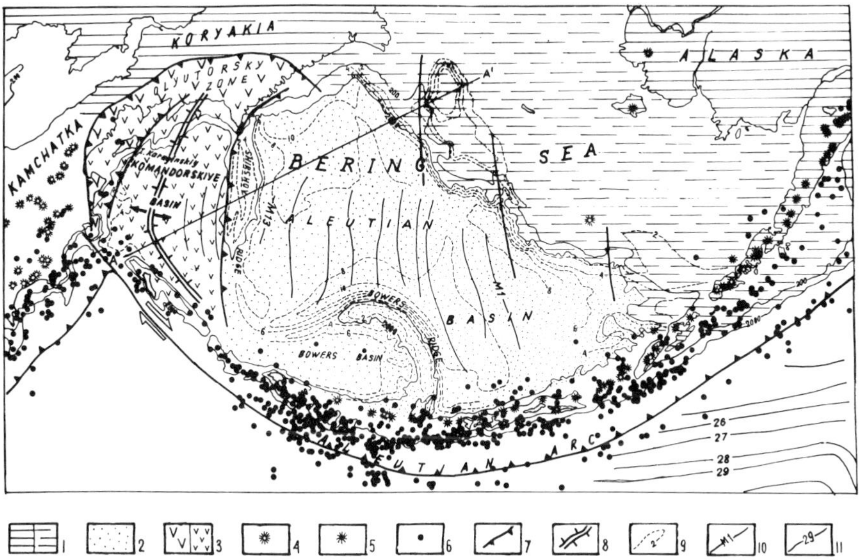

FIGURE 4. Structural-tectonic scheme of the Bering Sea region.

1 - areas of Mesozoic fold zone and pre-Albian accreationary prism: a) within continents; b) within continental shelf. 2 - pre-Albian oceanic plate; 3 - pre-Middle Miocene oceanic plate: a) within continent; b) within the sea. 4 - active volcanoes. 5 - extinct volcanoes. 6 - epicenters of earthquakes. 7 - subduction zones, 8 - axis of spreading. 9 - thickness of the sediments on the Aleutian plate and on the edge of the shelf (isolines are given in 2 km). 10 - magnit anomalies in the Bering Sea (according to A.Cooper). 11 - magnit anomalies in the Pacific plate.

BASIC FEATURES OF GEOLOGY OF THE SEA OF OKHOTSK

GNIBIDENKO, HELIOS S.(1), KHVEDCHUK, IGOR I.(2)
1. Sakhalin Complex Sci. Res. Institute, Yuzhno-Sakhalinsk, Sakhalin 693002, USSR. 2. Trust "Dalmorneftegazgeofizrazvedka", Yuzhno-Sakhalinsk, Sakhalin 693004, USSR.

ABSTRACT

The epi-Mesozoic Okhotsk platform occupies the central and northern parts of the Okhotsk Sea. The platform basement is composed of deformed geosynclinal formations, covering the stratigraphic range from Cretaceous up to pre-Cambrian inclusively. A slightly deformed cover levels the rough surface of the basement and is composed of Paleogene-Neogene terrigenic sedimentary formations, filling the system of structural-sedimentary basins.

The South Okhotsk Deep-Sea Basin occupies the southern part of the sea. This is an elementary basin, non-compensated by sediments which is in the back-arc of the Kuril geoanticlinal uplift. Two seismic units are distinguished within the sedimentary cover, the lower of which being, probably, composed of pelagic and the upper one of turbidity deposits covering as a whole the stratigraphic range from Upper Cretaceous to Holocene.

The development of a system of normal structural-sedimentary basins of the Okhotsk platform, which has begun in Paleogene, is associated with the crustal extension beginning from Paleogene in the direction of north-east and south-west, normal to the strike of the main epi-Mesozoic structural elements. The South-Okhotsk Deep-Sea Basin exists at any rate since Cretaceous time as a relict geosynclinal basin, separated from the Pacific as a result of appearance of the Kuril geoanticlinal uplift.

Proceedings of the 27th International
Geological Congress, Volume 23, pp. 19–41
ORIGIN AND HISTORY OF MARGINAL AND INLAND SEAS

INTRODUCTION

The shelf, continental slope and South-Okhotsk Deep-Sea Basin are the basic elements of the Okhotsk Sea relief (Fig.1).

The Okhotsk Sea shelf can be subdivided into the Near-Kamchatka, North-Okhotsk and Near-Sakhalin parts and also a plain borderland with uplifts, basins and troughs in a central part of the sea, which according to the depth (from 200 to 1000m) should be referred to the over-deepened shelf.

In a peripheral part of the South-Okhotsk Deep-Sea Basin a gradient of the continental slope is 5-10°, while on the Kuril Ridge slopes it reaches 20-30°. The slope is 3000 m high mainly. The continental slope rise borders the deepsea basin at depths of about 3000-3200 m like an apron of inclined (till to 1°) abyssal plain.

The South-Okhotsk Deep-Sea Basin floor is an abyssal plain extermely levelled at depths about 3200 m.

The Kurill Ridge and Kamchatka Peninsula separate the Okhotsk Sea from the Pacific Ocean and the Sakhalin-Hokkaido islands separates the Okhotsk and Japan Seas.

Recently the Okhotsk Sea is intensively studied and the geological and geophysical data are summarized in some papers (Gnibidenko, Tuezov, 1976; Burk, Gnibidenko, 1977; Gnibidenko, 1979; Belousov, 1981; Kornev, 1982; Gnibidenko Khvedchuk, 1982; Bikkenina, 1983). The aim of this paper is to review the Okhotsk Sea geology basin on the recent data.

The Okhotsk platform occupies the northern and central parts of the Sea of Okhotsk. The Sakhalin folded system borders it in the west. In the north it is bordered by the Okhotsk part of the Chukotsk-Katasian volcanic belt and in the east it is bordered by the Kamchatka folded system. In the south the platform is bordered by the South-Okhotsk Deep-Sea Basin (Fig.2).

THE OKHOTSK PLATFORM

According to the available data (Gnibidenko, 1979; Belousov, 1981; Kornev 1982) the basement and a cover (Fig. 3) are clearly distinguished in the Okhotsk platform structure.

BASEMENT

From the seismic data the roof of the basement is traced within the entire platform. It is characterized by seismic P-wave velocities of about 4.0-6.0 km/s, being separated from the sedimentary cover with an average velocity of about 2.0-3.5 km/s by a well pronounced unconformity. On vast areas the basement outcrops onto the śea floor from the sedimentary cover (Fig.4), which allows to obtain the data on its composition and extrapolate them to the troughs adjacent to the basement outcrops.

Dredging of these outcrops showed that basement rocks of the central part of the Okhotsk Sea forms a geosynclinal formation and is represented by sedimentary-volcanogeneous and igneous rocks: phyllites, greenschists, graywackes, aleurolites, sandstones, siliceous rocks, tuffs, tuff breccias of andesites, andesite-dacites, andesite-basalts, rhyolites, and also by quartz-diorites, granodiorites, granites, andesites, andesite-basalts, quartz pophyres and dacites.

The occurrence among the dredged material of metamorphic rocks (gneisses, granite-gneisses, quartzites, greenschists and phyllites) with developed cleavage testifies to the far advanced process of tectonic development and gives reason to believe that the basement, at any rate within the region of dredging, is a folded geosynclinal complex.

The unconformity between the sedimentary cover and the basement within the Okhotsk platform becomes more soundly pronounced towards Sakhalin, Kamchatka and to the northern coast of the Okhotsk Sea, where it can be correlated with the unconformity, separating the Cenozoic sedimentary cover from the basement, composed of Mesozoic and more ancient folded formations.

The conclusions as to the basement age of the Okhotsk platform are based on its tracing from the adjacent continental areas (North-East of the USSR, Kamchatka and Sakhalin) beneath the sedimentary cover, and also on the radiometric age determinations of the involved dredged rocks. K-Ar ages of the basement rocks dredged from the basement outcrops in the central part of the sea (Burk, Gnibidenko, 1977) appeared to be Upper Cretaceous and up to Middle Triassic. The age of granodiorites at the axial part of the Academy of Sciences of the USSR rise is 113-122 Ma (Upper part of Lower Cretaceous) and in St.Iona Island - 45-53 Ma (Gnibidenko, 1979).

Taking into account seismic wave velocities in the dredged rock samples (Gnibidenko, 1979) and also seismic wave velocities in the consolidated crust cover (Galperin, Kosminskaya, 1964; Geodekyan et al.,1980) one can conclude that the basement of the Institute of Oceanology Uplift (Fig.2) is no younger than Upper Cretaceous-Lower Paleogene. On the northern slope of the Academy of Sciences of the USSR Uplift the basement seems to be of Upper Cretaceous, while in its central part it is of Lower Cretaceous and more ancient. Thus, according to the available data one can believe that the basement of the Okhotsk platform includes folded geosynclinal formations, covering the stratigraphic range from Lower Paleogene-Cretaceous up to Upper-Middle Paleozoic and, probably, in some outcrops it is represented by geosynclinal formations of Paleozoic and pre-Cambrian ages.

It should be noted that slightly altered granodiorites, quartz-diorites, granites, and also basalts, andesites, rhyolites and their tuffs together with tuffogenic sandstones, aleurolites and argillites dredged from axial parts of the Academy of Sciences of the USSR and the Institute of Geology Uplifts and also from the Kashevarov Swell seem to build up Upper Cretaceous-Lower Paleogene postorogenic volcanogenic formations developed at the beginning of taphrogenesis, but before the formation of the terrigenic Upper Paleogene-Neogene cover.

COVER

The sedimentary cover overlaps and levels the basement

rough surface on the major part of the Okhotsk platform. The sedimentary cover thickness varies from 0.2-0.5 km within uplifts, increases up to 1-4 km, reaching more than 5-7 km (till 9 km) within the basement troughs (Fig.4). In most places of the Okhotsk platform the sedimentary cover is characterized by well reflected horizons traced over tens of kilometers and in some places according to the seismic records the basement can be subdivided into 2,3,4 or more units (Fig.3).

In central (the Academy of Sciences of the USSR Uplift) and western (Sakhalin) parts of the continental slope the sedimentary cover has been subjected to intensive erosion and dissected by canyons. One can find there submarine slides and the slope surface cut gently the sedimentary cover.

In structural-sedimentary basins deformations of the cover are represented by large gentle brachiform folds and faults. Only in some cases (East-Sakhalin basin, West-Kamchatka basin, the basin on the south slope of the Academy of Sciences of the USSR Uplift) the sedimentary cover has a relatively regular folded structure.

According to the dredging data and also the data of drilling in the East-Sakhalin and West-Kamchatka basins the sedimentary cover is represented by interbedded sandstones aleurolites and argillites, sometimes tuffogenic and diatomic, which were formed within the shelf conditions.

Taking into account dredging data and the correlation of sea and land cross-sections the stratigraphic range of the sedimentary cover is Upper Paleogene-Neogene, including Quaternary deposits of low thickness.

THE SOUTH-OKHOTSK DEEP-SEA BASIN

The South-Okhotsk Deep-Sea Basin (Fig.1,2 and 4) is a classical example of a basin, non-compensated by sediments locating in the back-arc of the island (Kuril) arc. An extremely levelled abyssal plain at a depth of about 3200-3300 m is characteristic of this basin. According to the crustal structure (suboceanic) and thickness (about 7-10 km) it is a structural element, not belonging to the Okhotsk platform, which is composed of continental and subcontinental crust about 20-30 km thick (Galperin, Kosminskaya, 1964).

BASEMENT

For lack of direct data on the composition of the basement, reasonable assumptions can be made only according to the indirect data on the composition and age of the rocks dredged along the border of the basin (Gnibidenko, 1979;Kornev,1982) and tracing of the basement (Fig.5,6) from the periphery of the basin towards its central part. The acoustic basement established in the basin from multchannel seismic profiling (Gnibidenko, Khvedchuk, 1982) coincides with the roof of layer 2 of the oceanic crust in the basin, having P-wave velocities of 5.1-5.2 km/s (Bikkenina, 1983) within the basin layer 2 (sedimentary-volcanogenic) has thickness of 1.0-1.5 km and is underlain by layer 3 (basaltic) with P-wave velocities of 6.4-6.8 km/sec (Fig.5) and thickness being about 3.5 km.

Taking into account the seismic P-wave velocities within the roof of layer 2 (5.1-5.2 km/s; the deep-sea basement) and its thickness (1.0-1.5 km), one can assume that its composition is similar to the composition of the oceanic crust layer 2 and thus seems to be represented by the basalts intercalated with sedimentary rocks (argillites). Layer 2 thickness becomes 2-3 times higher, its roof being composed of Mesozoic and Mesozoic-Upper Paleozoic (Sakhalin) eugeosynclinal deposits within the folded border of the basin.

Dredging of separate basement outcrops along the Kuril geoanticlinal uplift (north-westward of Iturup Island; B.I.Vasiliev, 1982, personal communication; Fig.1) showed that basic volcanites (basalts, diabases and their tuffs) which seem to intercalate with aleurolites, schists and sandstones are in fact involved into the composition of the basin basement. These basement outcrops seem to be overlapped by lavas and tuffs of acid effusives and tuffogenic-sedimentary rocks and broken by granodiorites.

If to believe that granodiorites dredged from submarine outcrops of the basement northward of Urup and Simushir

Islands (Fig.1) relate to the crustal layer 2 at the base of the Kuril geoanticlinal uplift, and have K-Ar ages of 155-220 Ma (Gnibidenko, Khvedchuk, 1982) one should suppose that Jurassic and more ancient formations can be involved into the basement composition of the Okhotsk Deep-Sea Basin.

COVER

The sedimentary cover of the Okhotsk Deep-Sea Basin has maximum thickness of about 4.0 km and is clearly subdivided into two units (Fig.6): the upper stratified layer with thickness of about 1.0 km and P-wave velocities about 1.5-2.3 km/s and the lower acoustically "transparent" layer with thickness of about 3.0 km and P-wave velocities about 2.2-4.5 km/s, which, in its turn, according to the seismic refraction data (Fig.5), by the boundary with P-wave velocity of 4.3 km/s is subdivided into the upper and lower parts (Popov, Anosov, 1978; Bikkenina, 1983).

It is obvious that the upper layer is the interbedding of turbidites with volcanogenic intercalations (ashes), probably, covering the stratigraphic range from Upper Miocene to Pliocene-Quaternary. On the basement outcrops northwestward of Iturup Island the dredged sedimentary cover rocks is usually represented by tuffogenic sandstones, diatomic aleurolites and argillites (B.I.Vasiliev, 1982; personal communication). The lower ("transparent") layer probably is composed by clays and argillites. If the sedimentation rate for the "transparent" layer is supposed to be 20 m/mln years the sedimentary cover formation should be reffered to pre-Upper Cretaceous, even without taking account of consolidation of sediments after deposition. The lower layer thickness being supposed to be approximately half decreased as a result of consolidation, the beginning of its formation can be referred to Lower Mesozoic or Upper Paleozoic.

The sedimentary cover of the basin is deformed slightly (fig.6), mainly by faults. The upper (stratified) layer and upper part of the acoustically "transparent" layer onlap the Mesozoic basement over the basin periphery (Fig. 6). However the onlap of the lower part acoustically "transparent" layer (with velocity of 4.0-4.3 km/s; Bikkenina, 1983) on the basement is not so obvious and it seems that this part of "transparent" layer is transferred into the basement of the Kuril geoanticlinal uplift and the continental slope (with P-wave velocities about 4.8-5.5 km/s; Fig.5 and 6) involving into the folding and faulting deformations.

DISCUSSION: GEOLOGICAL DEVELOPMENT OF THE OKHOTSK SEA REGION

From the above data one can conclude that heteroaged folded geosynclinal formations compose the basement of the Okhotsk platform, the sedimentary cover of which has begun to form only in Paleogene and seems to include Upper Paleogene and Neogene-Quaternary terrigenic deposits.

The Okhotsk arch (Fig.2) probably is the most ancient structural element of the Okhotsk platform. Eugeosynclinal system was formed in Early Proterozoic (Grinberg, 1968) or Archean (about 4 Ga; Korolkov et al., 1974) at the place of the Okhotsk arch if to judge accor ding to its northern block - the Okhotsk block of early deformations (Gnibidenko, 1969). A mature geosynclinal system had been formed there already in Archean and Early Proterozoic. By Upper Proterozoic it had been transformed into a rigid massif within which miogeosynclinal and later orogenic regimes existed, if to refer to the data for the part of the arch within the land (Gavrilov, 1965; Veldyaksov et al., 1970; Chikov, 1970), since Middle Proterozoic till Lower Paleogene. Probably since Middle Paleogene the Okhotsk arch was above the sea level till Pliocene and then subsided and was overlapped by a thin cover of Pliocene-Quaternary deposits.

According to the available data a geosynclinal system began to develop at the place of the Okhotsk folded system, locating northward of the Okhotsk arch to the adjacent land (Drabkin, v.30,1970; Bely, Kotlyar, 1975), in pre-Cambrian. Already since Upper Paleozoic this region is characterized by miogeosynclinal regime replaced in Triassic-Jurassic by andesitic volcanism in axial parts of geoanticlinal uplifts and then by Cretaceous-Paleogene volcanism (Osipov, 1975) within the Okhotsk volcanic belt. At the end of Cretaceous-Paleogene taphrogenesis has led to large subsidences within the Okhotsk platform, which, probably, as pothumous movements first of all was in the areas of geosynclinal basins, at places of which by that time synclinoria had been formed. Thus, there had been formed the Severny and West Okhotsk troughs compensated then by terrigenic Paleogene and Neogene deposits.

Miogeoanticlinal uplift seems to be formed by Upper Cambrian-Ordovician formations (from the data on the Yurovsky block; Korolkov, 1972) at the place of the Kashevarov Swell, located along the south-western edge of the Okhotsk arch (Fig.2). However, eugeosynclinal regime seemed to be at the major part of the swell (within the sea area) during Paleozoic and Mesozoic till Eocene. By that time the swell was an extended island arc which died by Neogene and then was eroded and its southern part was subsided to 0.5-1 km and was overlapped by Upper Pliocene-Quaternary deposits.

Ensymatic geosyncline (Shahkin, 1970) began to develop in pre-Cambrian at the place of Mongolo-Okhotsk folded system separated the Aldan shield and the Okhotsk arch. This geosyncline had been closed in Middle Mesozoic (Nagibina, 1969). A volcanic belt was formed there during Cretaceous. In Paleogene the vast area of the former geosyncline was land eroded and, probably, subsided in Pliocene and overlapped by the sedimentary cover.

Ensymatic geosynclines (Gnibidenko et al.1974; Gnibidenko et al., 1975) began to develop at the place of the Western Kamchatka and Salkalin-Hokkaido folded systems in Upper or Middle Paleozoic and in Paleogene mature geoanticlinal uplifts with volcanogenic-molassa type of sedimentation were formed at their places. Folded deformations spread from the axial zones of those systems towards their periphery from Upper Cretaceous till Upper Miocene and Post-Pliocene folding deformations continue as posthumous movements in Quaternary (Soloviev, 1972). Quaternary deformations can also be supposed in the TINRO basin.

The process of tectonic expanding of the Okhotsk platform in the northern part pf the Okhotsk Sea (Alexeyev et al., 1975) seems to continue at present. Glacioeustatic regressions to 100-110 m during Quaternary resulted in drainage of vast areas in the north of the Okhotsk Sea (Kulakov, 1973). However, glacioeustatic regressions of the ocean to 1000 m seem to be improbable (Gilshe, 1974) and therefore subsidences of the platform basement below 150-200 m of the recent ocean level should be connected with tectonic reasons.

Taking into account the ideas on the age of the Okhotsk platform basement and cover and also on the structural relation of its central part to folded land areas one can conclude that the geosynclinal system began to develop in the central and southern parts of the platform, probably, in Upper and Middle Paleozoic and by the beginning of Paleogene a system of eugeoanticlinal and miogeoanticlinal uplifts was formed. Those uplifts outcropped as island arcs and miogeosynclinal troughs, non-compensated by sediments, which separated them, occupying the area from the southern edge of the Okhotsk arch up to the recent continental slope. The ceasing of volcanism and denudation at the end of Paleogene and in Miocene resulted in peneplain and then the central and southern part of the Okhotsk platform subsided to about 1000 m. However, this subsidence seemed to be gradual enough and the regions of the Institute of Oceanology and the Academy of Sciences of the USSR Uplifts were involved into the subsidence only at the end of Pilocene, as they were overlapped by a thin cover of Upper Pliocene-Quaternary deposits.

The Cenozoic movements along deep faults within the Okhotsk platform mainly of the north-western and sub-latitudinal strike stipulated the formation of a rift system in the Okhotsk platform basement (Gnibidenko, 1976). From seismic profiling data one can conclude that Pliocene-Quaternary deposits are also cut by fault deformations.

The South-Okhotsk Deep-Sea Basin exists as a stable geomorphological element with almost recent boundaries, probably, since pre-Upper Cretaceous or earlier time. The basin borders seemed to be formed when geosynclinal systems began to develop at the places of the folded system of the Academy of Sciences of the USSR, Sakhalin-Hokkaido and Kuril Uplifts, that is in Early Mesozoic or Upper Paleozoic. The development of those systems into the mature geoanticlinal uplifts of island arcs in Mesozoic and Cenozoic resulted, probably, in the fact that already at the end of Cretaceous-Lower Paleogene this basin was a semiclosed basin, non-compensated by sediments. Probably, a land areas, stipulated a turbidity character of sedimentation for the upper part of the sedimentary cover of the basin, was formed at the place of those geosynclinal systems, bordering the basin by the beginning of Neogene.

Non-deformed sedimentary cover structure with normal faults over the periphery of the basin testify to the fact that, probably, since pre-Cretaceous there had been stable conditions relative to the abyssal sedimentation.

A Quaternary regime of sedimentation in the basin is mainly inherited and is characterized by terrigenic sedimentation with decrease of mean granulometric compostion of sediments towards its central part (Ilyev et al., 1979) where aleurite oozes are accumulated, probably in the form of distal turbidites. Thus, the recent South-Okhotsk Deep-Sea Basin is a miogeosynclinal basin, non-compensated by sediments with heat flow, exceeded normal one twice (Ehara, 1979), which seems to be stipulated by the relatively heated upper mantle.

Heat flow anomalies in the Okhotsk Sea (Gnibidenko, Khvedchuk, 1982) point to the differences in thermodynamic conditions at compared dephts (Tikhomirov, 1970) and in active tectonic processes in the Earth's crust and upper mantle. Considerable isostatic unbalance of the central sea regions (Gainanov et al., 1971) seems to testify to the continued vertical movements of crustal and mantle blocks. The evalutions of isostasy of the Okhotsk Sea region (Gainanov et al., 1974) give reasons to assert that the Okhotsk platform and the South-Okhotsk Deep-Sea Basin are in the state of differentiated unbalance. Thus, central parts of the Okhotsk platform are characterized by banded alternation of areas of overcompensation and undercompensation, which together with the anomolous heat flow, probably, testify to the continued processes of the upper mantle material differentiation. Positive isostatic anomalies of the South-Okhotsk Deep-Sea Basin and especially intensive positive isostatic anomalies of the Greater (volcanic) Kuril Islands Ridge, probably, also testify to the processes of the upper mantle material differentiation, which support those regions in the state of undercompensation. The alternation of the upper mantle layers with increased (8.0-8.4 km/s) and decreased (6.8-7.0 km/s) P-wave velocities beneath the South-Okhotsk Deep-Sea Basin up to the depth of about 30 km (Starshinova, 1980) seems to stipulate such a differentiation of the mantle material, which results in its partial melting.

ACKNOWLEDGEMENTS

The authors thank Mrs. Zarema S.Polyakova for translating the paper into English and Miss Valentina B. Makarova for typing the manuscript.

REFERENCES

Alexeyev M.D, Onukhov F.S, Ufimtsev G.F. 1975. Morphotectonics of the floor of the Okhotsk Sea northern part. Proc. Acad. Sci. USSR, 903-905 (in Russian).

Belousov V.V. (ed.) 1981. The structure of the Okhotsk Sea. Nauka, Moscow, 176 pp. (in Russian).

Bely V.F, Kotlyar I.N. 1975. New data on geology of the Piyagin peninsula western part. In: N.A.Shilo (ed.), The Data on Geology and Mineral Resources of the north-east of the USSR. Proc. North-East Complex Sci. Res. Inst., 22, 74-85 (in Russian).

Bikkenina S.K. 1983. Peculiarities of the deep structure of the Southern Okhotsk Sea according to new seismic data Synopsis Thesis. Khabarovsk, 19 pp. (in Russian).

Burk C.A. and Gnibidenko H.S. 1977. The structure and age of acoustic basement in the Okhotsk Sea. In: M.Talwani and W.C. Pittman III (eds.), Island Arcs, Deep Sea Trenches and Back-Arc Basins. Maurice Ewing Ser., 1, 451-466.

Chikov B.M. 1970. Tectonics of the Okhotsk Medial Massif.

Nauka, Moscow, 152 pp. (in Russian).
Drabkin I.E. (ed.), 1970. Northeast of the USSR. Book 1 and 2. Geology of the USSR, Vol. 30. Nedra, Moscow, 1084 pp. (in Russian).
Ehara S. 1978. The heat flow in Hokkaido-Okhotsk sea region and its tectonics implications. In: S.L.Soloviev (ed.), Structure and Geodynamics of North-West Pacific Lithosphere According to Geophysical Data. Far East Center Academy of Sciences of the USSR, Vladivostok, pp.86-98 (in Russian).
Gainanov A.G., Isaev E.N.,Stroev P.A. and Ushakov S.A. 1971. Isostasy and crustal structure of the Okhotsk Sea region. Geophys. Bull. Intergovernment Geophys. Comm. Acad.Sci.USSR, 2, 37-43 (in Russian).
Gainanov A.G., Pavlov Yu.A., Stroev P.A., Syshev P.M. and Touesov I.K. 1974. Anomalous Gravity Fields of the Far East Marginal Seas and the Adjacent Part of the Pacific Ocean. Nauka, Novosibirsk, 108 pp. (in Russian).
Galperin E.G. and Kosminskaya I.P. (eds) 1964. Crustal Structure in the Asia-to-Pacific Transition Zone. Nauka, Moscow, 307 pp. (in Russian).
Gavrikov S.I. 1965 On the history of geological development of the Okhotsk massif. Sov. Geol., 2, 71-81 (in Russian).
Geodekyan A.A., Neprochnov Yu.P., Elnikov I.N. and Yaroshevskaya G.A. 1980. The results of the seismic studies on the Academy of Sciences of the USSR Rise in Okhotsk Sea. Acad. Sci. USSR, 250, 919-923. (in Russian).
Gilshe A.V. 1974. Change of the sea level in Pleistocene and Holocene. In: V.F.Agranat (ed.), Quaternary glaciation of the Earth. Mir, Moscow, pp. 186-221 (in Russian).
Gnibidenko H.S. 1969. On the tectonics of the Okhotsk massif of early deformations. In: S.S.Zimin (ed.), Petrology of igneous and metamorphic rocks. Academy of Sciences of the USSR, Vladivostok, pp.73-81 (in Russian).
Gnibidenko H.S. 1976. On the rift system of the Okhotsk Sea floor. Proc.Acad. Sci. USSR, 229, 163-165 (in Russian).
Gnibidenkov H.S. 1979. The Tectonics of the Far East Marginal Seas. Nauka, Moscow, 161 pp. (in Russian).
Gnibidenko H.S. and Touezov I.K. 1976. The Okhotsk Sea.

In: H.S.Gnibidenko (ed.), The structure of the earth's crust and upper mantle in the ransition zone between Asia and Pacific. Nauka, Novosibirsk, pp.99-110 (in Russian).

Gnibidenko H.S., Gorbachev S.Z., Lebedev M.M and Marakhanov V.I. 1974. Geology and deep structure of Kamchatka Peninsula. Pac. Geol., 7, 1-32.

Gnibidenko H.S., Saito K., Zassu S. and Ozima M. 1975. On the age and genesis of the rocks from the basement of the Sakhalin Island. In: H.S.Gnibidenko (ed.), Natural Resources of the Sakhalin, its development and using. Sakhalin Branch of the Geographical Society of the USSR, Yuzhno-Sakhalins, pp. 93-103 (in Russian).

Gnibidenko H.S. and Khvedchuk I.I. 1982. The tectonics of the Okhotsk Sea. Marine Geology, 50, 155-198.

Grinberg G.A. 1968. The Okhotsk Pre-Cambrian Massif. Nauka, Moscow, 187 pp. (in Russian).

Iliev A.I., Voronova V.A., Zakharova M.A., Nesterova O.N., Tarakanova L.I., Scheremetyeva G.N. and Schustov L.N. 1979. The bottom sediments of the Southern Okhotsk Sea. Nauka, Moscow, 148 pp. (in Russian).

Kornev O.S. (ed.) 1982. Geological Structure of the Okhotsk Sea Region. Academy of Sciences of the USSR, Vladivostok, 113 pp. (in Russian).

Korolkov V.G.1972. On stratigraphy of the Okhotsk region Lower Ordovician deposits. In: N.E.Drabkin (ed.), Data on Geology and Mineral Resources of the Northeast of the USSR. Proc. North-East Geol. Surv. USSR, 20, 31-33 (in Russian).

Korolkov V.G., Rudnik V.A. and Sobotovich E.A. 1974. On Late Paleozoic-Early Archean age of the most ancient rocks of the Okhotsk Medial Massif. Proc. Acad. Sci. USSR, 219, 1441-1444. (in Russian).

Kulakov A.P. 1973. Quaternary Coastal Lines of the Okhotsk and Japan Seas. Nauka, Novosibirsk, 188 pp. (in Russian).

Nagibina M.S. 1969. Stratigraphy and Formations of the Mongolo-Okhotsk Belt. Nauka, Moscow, 400 pp. (in Russian)

Osipov A.P. 1975. Magmatic formations of the Okhotsk-Chukotsk volcanic belt. Geol. Explor. , 3, 16-22 (in Russian).

Popov A.A. and Anosov G.I. 1978. New data on crustal structure of the Kuril Basin. Proc.Acad.Sci. USSR, 240, 166-168 (in Russian).

Shashkin K.S. 1970. Structure-formational zones of Mongolo-Okhotsk belt and some features of their development. In: I.N.Govorov (ed.), Some Problems Related with Geochemistry and Metallogeny of Northwestern Pacific Belt. Academy of Sciences of the USSR, Vladivostok, pp.31-33 (in Russian).

Soloviev V.V. 1972. Recent tectonics of the Sakhalin Island. In: A.S.Khomentovsky and S.M.Tseitlin (eds.), Problems of Study of a Quaternary Period. Nauka, Moscow, pp. 322-326 (in Russian).

Starshinova E.A. 1980. Crustal and upper mantle heterogeneity of the Okhotsk Sea. Proc.Acad.Sci.USSR, 255, 1339-1343 (in Russian).

Tikhomirov V.M. 1970. Thermodynamic conditions within the Earth's crust and upper mantle of the Okhotsk Sea, Kuril Islands and near-Kuril parts of the Pacific Ocean. In: H.S.Gnibidenko (ed), Geology and Geophysics of the Pacific Belt. Proc.Sakhalin Complex Sci. Res. Inst., 25, 23-33 (in Russian).

Veldyaksov F.F., Ivanov V.A., Peskov E.G. and Ryabov A.V. 1970. Main features of tectonics and development of the Okhotsk massif. In: K.V.Bogolepov (ed.), Tectonics of Siberia, Vol.4, Nauka, Moscow, pp.82-91 (in Russian).

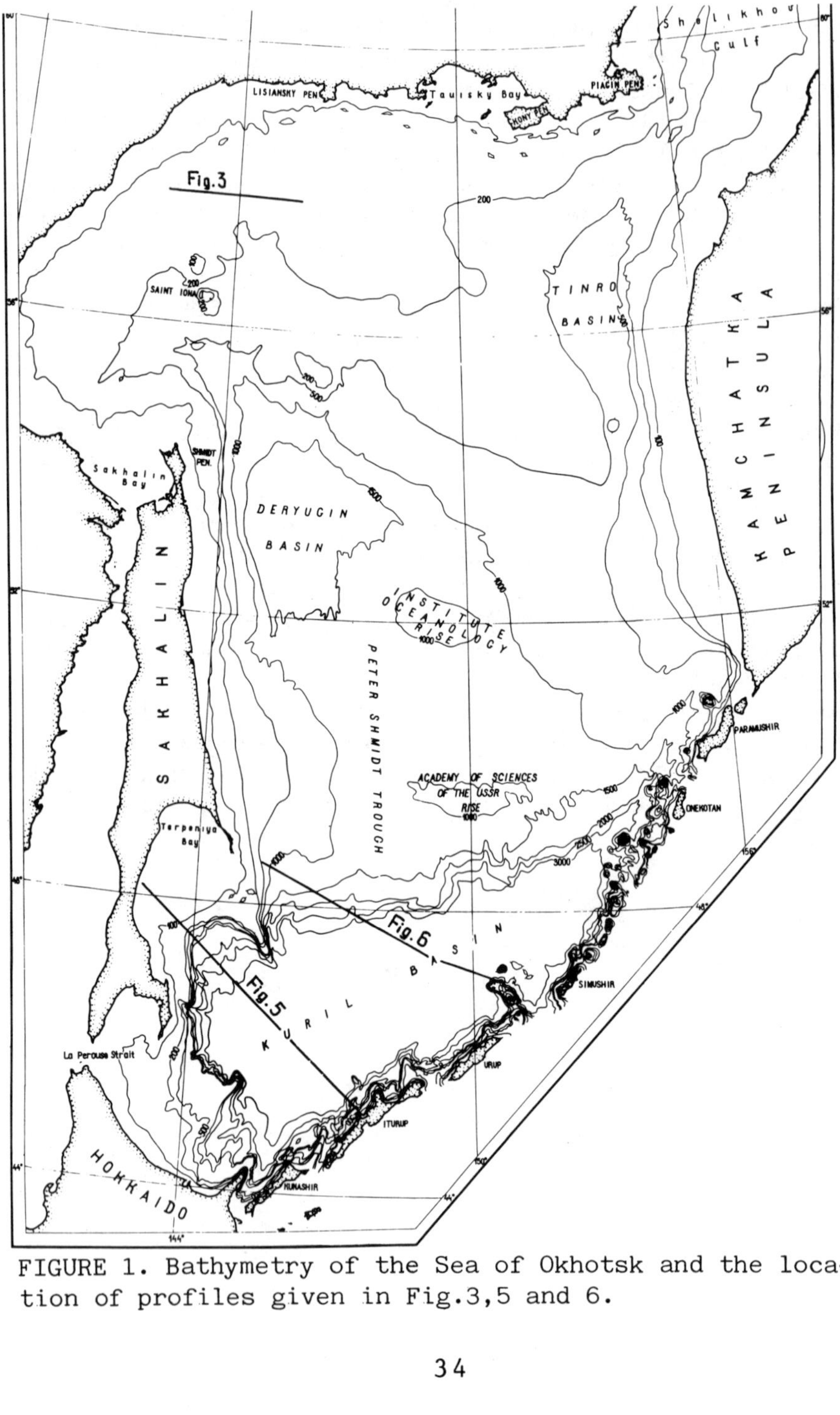

FIGURE 1. Bathymetry of the Sea of Okhotsk and the location of profiles given in Fig.3,5 and 6.

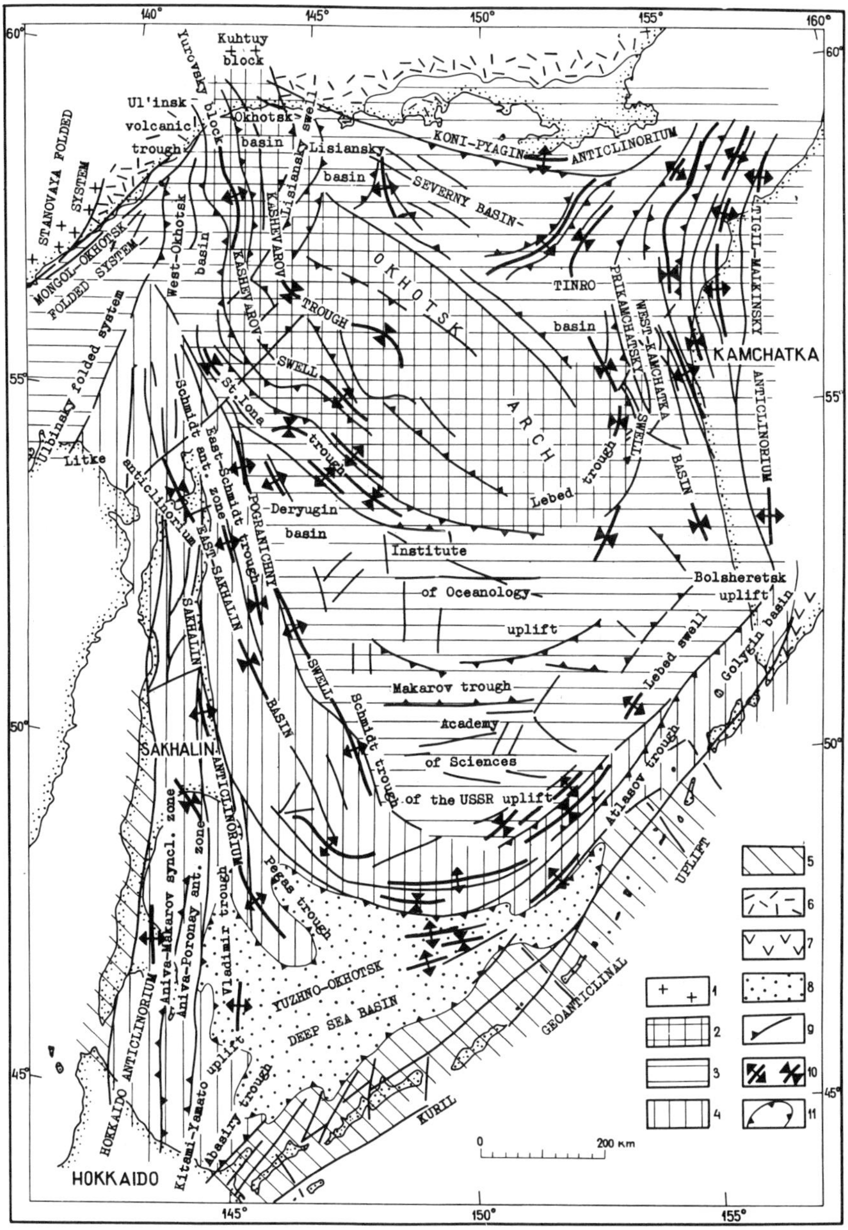

Kuhtuy block
Yurovsky block
Ul'insk volcanic trough
STANOVAYA FOLDED SYSTEM
MONGOL-OKHOTSK FOLDED SYSTEM
Ulbinsky folded system
West-Okhotsk basin
Okhotsk basin
Lisiansky swell
Lisiansky basin
KASHEVAROV TROUGH
KASHEVAROV SWELL
KONI-PYAGIN ANTICLINORIUM
SEVERNY BASIN
OKHOTSK ARCH
TINRO basin
PRIKAMCHATSKY SWELL
WEST-KAMCHATKA BASIN
TIGIL-MALKINSKY ANTICLINORIUM
KAMCHATKA
St. Iona trough
Schmidt ant. zone
East-Schmidt trough
EAST-SAKHALIN ANTICLINORIUM
POGRANICHNY BASIN
Litke
anticlinorium
Deryugin basin
Institute of Oceanology uplift
Lebed trough
Bolsheretsk uplift
Golygin basin
Lebed swell
Makarov trough
Academy of Sciences of the USSR uplift
Atlasov trough
SWELL
Schmidt trough
SAKHALIN
Pegas trough
Aniva-Makarov syncl. zone
Aniva-Poronay ant. zone
Vladimir trough
Kitami-Yamato uplift
Abasiry trough
YUZHNO-OKHOTSK DEEP SEA BASIN
KURIL GEOANTICLINAL UPLIFT
HOKKAIDO ANTICLINORIUM
HOKKAIDO
200 km

FIGURE 2. The general tectonics of the Okhotsk Sea region. Precambrian deformed and metamorphosed complexes (1) of the Stanovaya folded system and the Kuhtuy Block. Paleozoic, Mesozoic and Cenozoic folded complexes (2-5): 2 - deformed in the late Paleozoic with the beginning of geosynclinal sedimentation in pre-Cambrian and early Paleozoic, 3 - deformed in the Mesozoic with beginning of geosynclinal sedimentation in early Paleozoic and late pre-Cambrian, 4 - deformed in the late Mesozoic and early Paleogene with the beginning of geosynclinal sedimentation in early Paleozoic, 5 - deformed in the late Neogene (deformations continue to present) with the beginning of geosynclinal sedimentation in late Paleozoic and early Mesozoic, 6 - Okhotsk postgeosynclinal Cretaceous and Cenozoic volcanic belt, 7 - East-Kamchatka Neogene-Quaternary postgeosynclinal volcanic belt, 8 - Neogene-Quaternary turbidity unit of the South-Okhotsk Deep-Sea Basin, 9 - main faults with arrows to deep side, 10 - horstanticlinoria (a) and graben-synclinoria (b), 11 - boundary of the South-Okhotsk Deep-Sea Basin.

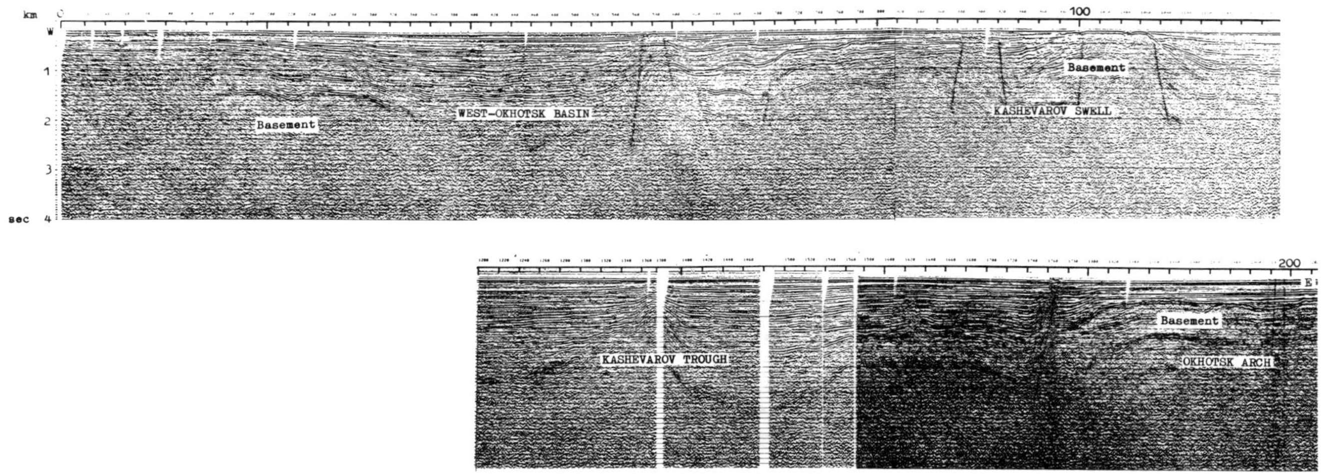

FIGURE 3. Multichannel seismic profiling cross-section in the north-western part of the Okhotsk platform. For location see Fig.1.

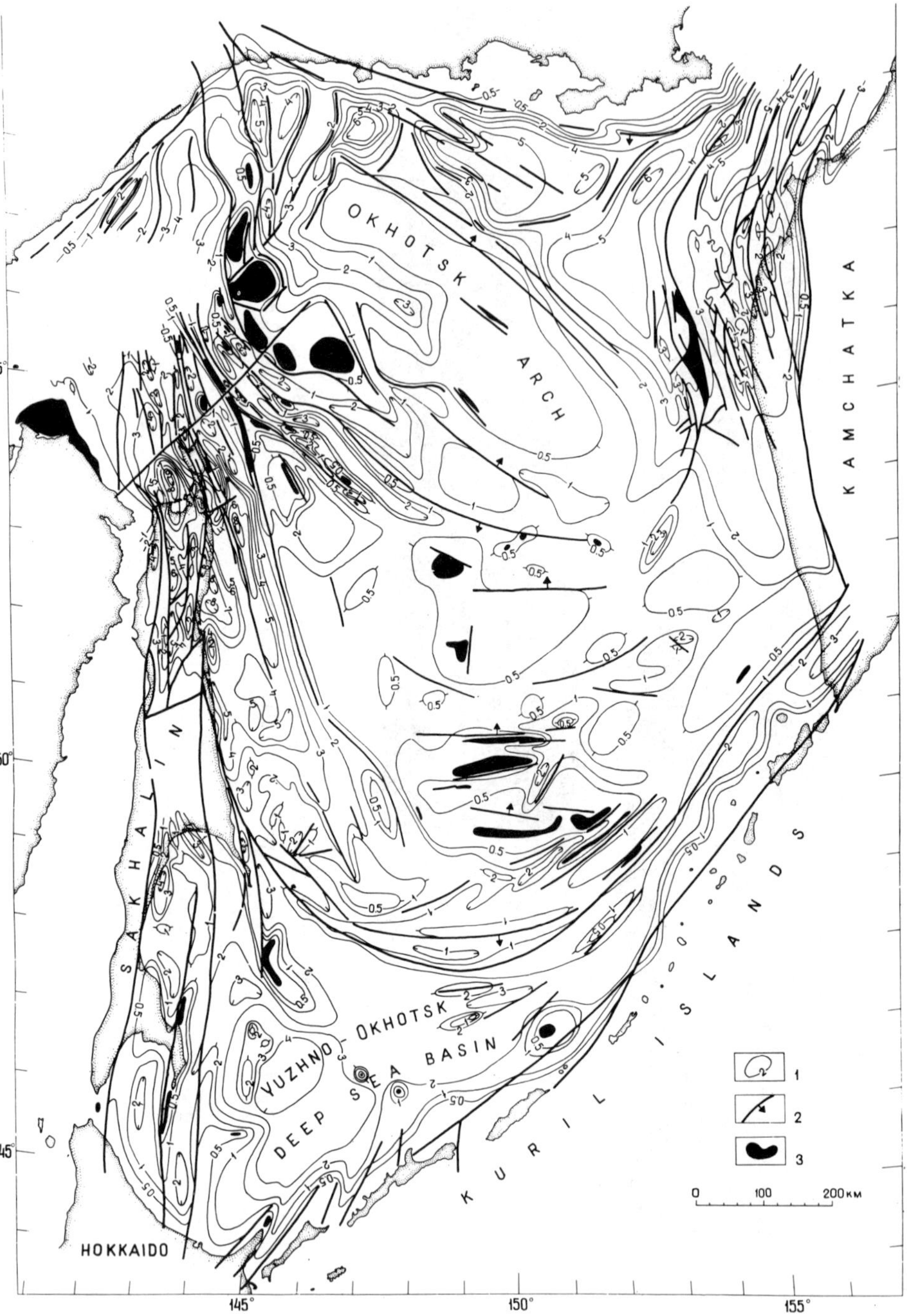

OKHOTSK ARCH
KAMCHATKA
SAKHALIN
YUZHNO - OKHOTSK DEEP SEA BASIN
KURIL ISLANDS
HOKKAIDO
1
2
3
0 100 200 KM
55°
50°
45°
145°
150°
155°

FIGURE 4. Isopach map of the Okhotsk Sea.
1 - isopachs of the Cenozoic sedimentary cover with seismic P-wave velocity ranging from 1.7-4.5 km/s; 2 - main faults with arrows to deep side; 3 - outcrops to the sea bottom of the acoustic basement are represented by the geosynclinal seformed pre-Cenozoic sedimentary, igneous and metamorphic rocks.

FIGURE 5. Seismic cross-section along the profile Sakhalin -Iturup Island across the South-Okhotsk Deep-Sea Basin. (Bikenina, 1983; for location see Fig.1).
1 - refracted boundaries compiled by reversed (a) and single (b) travel time curves; 2 - single determinations of the depths of refracted boundaries; 3 - places of interpolation; 4 - reflected boundaries according to the refracted wave method; 5 - depths to the acoustic basement along the single-channel seismic profiles with intersection of it with refracted method profiles; 6 - reflected boundaries according to multichannel seismic profiling; 7 - faults; 8 - boundaries from DSS data (Galperin and Kosminskaya, 1964); 9 - the acoustic basement roof according to multichannel seismic profiling; 10 - velocity cross-section of the upper mantle according (Starshinova, 1980); 11 - depths to seismic boundaries on the profiles obtained by refracted method and DSS in places of intersection with the considered profiles.

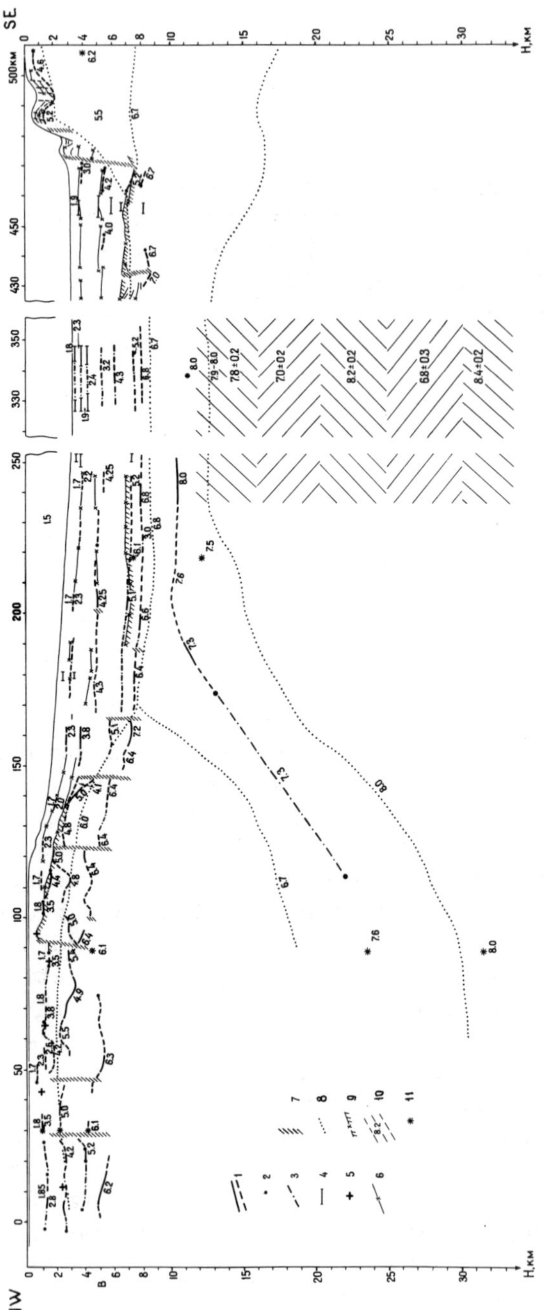
NW
SE
H,км
500км

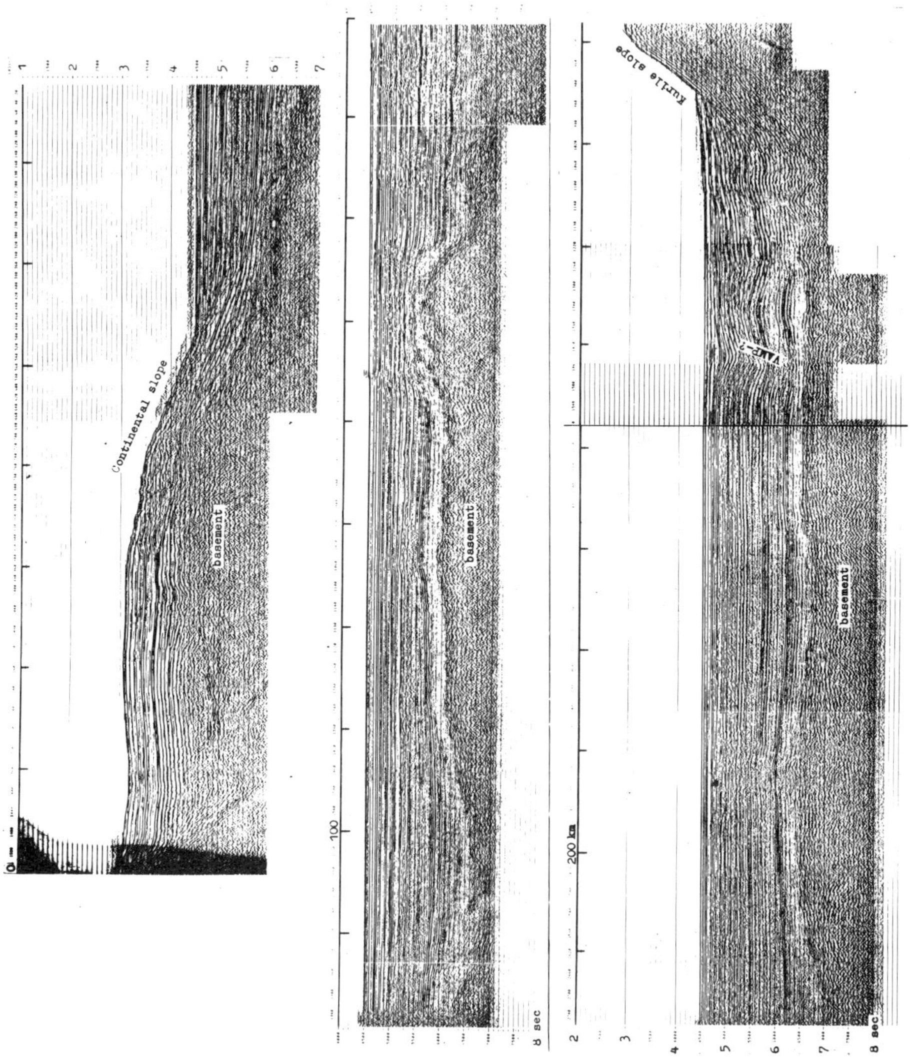

FIGURE 6. Multichannel seismic profiling cross-section (24 channels) across the South-Okhotsk Deep-Sea Basin. For locations see Fig. 1.

SEDIMENTARY BASINS OF THE JAPAN SEA

ISHIWADA, Yasufumi, Japanese Association for Petroleum Technology, Keidanren Kaikan, 1-9-4, Ohtemachi, Chiyoda, Tokyo 100,
HONZA, Eiichi and TAMAKI, Kensaku, Geological Survey of Japan, Higashi, Yatabe, Ibaraki 305, Japan

ABSTRACT

Numerical sedimentary basins are observed in the Japan Sea. Development of the sedimentary basins is closely related to tectonics of the Japan Sea since its initiation.

We have summarized the distribution of the sedimentary basins in the Japan Sea based on single and multi-channel seismic reflection data. Some of them are deep-sea sedimentary basins which are presently under active sedimentation. Others are developed along the coninental shelf of the Japanese Islands, and almost of them are not presently under active sedimentation. The thickest sediment deposition in the Japan Sea is not observed in the present deep-sea basin but in the sedimentary basins on the continental shelf along Japan, with the thickness exceeding 5.0 seconds in two-way travel time.

They were formed in associated with tensional and subsidence tectonics during Early Miocene to Late Miocene and succeeding E-W compressional tectonics since Pliocene to Present have caused uplift of sedimentary basins along the eastern margin of the Japan Sea which are presently on the continetal shelves as well as onshore area.

INTRODUCTION

Proceedings of the 27th International
Geological Congress, Volume 23, pp. 43–65
ORIGIN AND HISTORY OF MARGINAL AND INLAND SEAS

Several models or hypotheses are proposed for the origin of marginal seas. Among these models, there is a concept that the marginal sea is formed on the relation to the fromation of arc. In this concept, marginal sea is related to the suvduction of oceanic plate under the arc, and is counted as one of the fundamental elements which constitute arc (Honza, 1981).

The Japan Sea is categorized as one of the inactrive backarc basins (Karig, 1973), which is interpreted from relatively high heat flow, thick sediemts in the depocenter and possible older age's identification of magnetic lineations in the basins, although magnetic lineations are less clear in the Japan Sea (Isezaki, 1976). This magnetic feature is common in backarc basins.

Free-air gravity anomalies are 10 to 20 milligals and Bouguer gravity anomalies are lower than that in the Pacific Ocean (Ishihara, in press; Ludwig et al., 1975; Tomoda, 1973).

The Japan Basin has an oceanic plate with 30 km thick (Yoshii, 1972; Abe and Kanamori, 1970) in which the oceanic crust is 9 km thick with horizontal layers of the layer 3 (6.7 - 7.0 km/sec) with 5 km thick, the layer 2 (4.8 - 6.2 km/sec) with 2 km thick and sedimentary layers with 2.0 km thick in average. The Yamato Basin has an oceanic crust with 14 km thick in which the layer 3 is 9 km thick, the layer 2 with 2 km thick and sedimentary layers with less than 1.5 km (Ludwig et al., 1975; Murauchi, 1972; Tuezov, 1971).

Almost of ridges or rises in the Japan Sea consist of continental materials. They dominaltly consist of granitic rocks and others such as basaltic and sedimentary rocks (Tamaki et al., 1981; Honza ed., 1979; Iwabuchi, 1968).

The origin of the Japan Sea has been still controversial, although there have been voluminous data obrained in the Japan Sea (e.g., Gnibidenko, 1979; Soloviev et al., 1977; Hilde and Wageman, 1973). In this paper, we summarize sedimentary structure and stratigraphy of the Japan Sea on the basis of our single and multi-channel seismic reflection data, and briefly discuss its origin in relation to arc history.

MORPHOLOGY

The Japan Sea consists of basins, ridges and rises (Fig. 1). There are secveral basins, i.e., the Japan Basin with 3,500 m deep, The Yamato Basin with 3,000 m deep and the Tsushima Basin with 2,500 m deep. The Tartary Trough borders northern margin of the Japan Basin and deepens toward south.

The Yamato Ridge consists of a few banks of the Yamato, The Kitayamato and the Takuyo Banks. The Kitayamato Trough intervenes between the Yamato and the Kitayamato Banks. The Kitayamato Bank is neighbouring to the Yamato Bank to the north and to the Oki Spur to the south. The Korean Plateau divided into two of the north and the south is intercalated by a branch of the Japan Basin.

The Sado and the Okushiri Ridges develop along the coast of the northern Japan from Sado Island to the Tertary Trough.

A few deep-sea channels are observed along the coasts of the Japan Sea. The Toyama deep-sea channel starts from Toyama Bay and ends at the Japan Basin through the northern Yamato Basin. Other deep-sea channels such as the Tartary, Tsushima and Genzan deep-sea channels are less clear to trace, and have a discontinous stream.

STRATIGRAPY AND STRUCTURE

Sedimentary layers resulted from single channel and multi-channel seismic reflection measuremetns are correlated to geological ages based on the sampled results of dredges, corings and DSDP drillings in the Japan Basin (Site 301), in the Yamato Basin (Site 299) and in the northern margin of the Yamato Ridge (Site 302).

Deep-Sea Basin

The Japan and the Yamato Basins have several similarities and several interesting differences. One of the most striking difference occurs in the depth to acoustic basement which exceed 7.0 second (5,500m) below the sea level in the Japan Basin, but is not

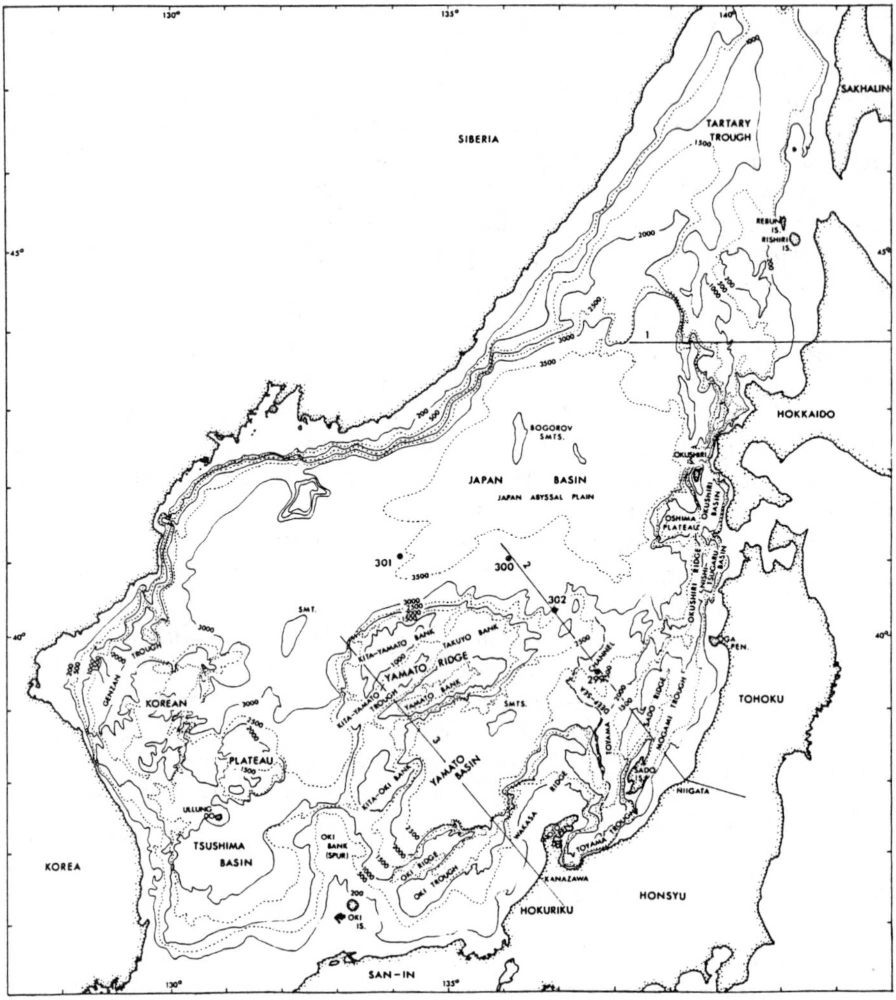

FIGURE 1. Physiography of the Japan Sea.
Bathymetry is based on the Hydrographic Office of Japan, Chert NO. 6301.

greater than 5.5 seconds (4,500m) in the Yamato Basin. Another notable difference is in sediments thickness; maximum thickness in the Japan Basin exceeds 2.0 seconds (2,000m) whereas in the Yamato Basin only 1.6 seconds (1,600m). The acoustic pattern of the sediments in both basins is similar (Fig. 2). In general the sediments can be divided acoustically into two units; an upper stratified layer and a lower less stratified layer. The difference in total sediment thickness of the two basins depend on variations in the thickness of the lower unit (Sato, 1971; Murauchi, 1972; Hilde and Wageman, 1973; Ludwig et al., 1975; Honza et al., 1977; Tamaki et al., 1979). A few distinct reflectors are distinguished in both basins, the uppermost reflector (reflector a) in the upper unit is approximately correlated to the boundary between Pleistocene and Pliocene from the DSDP drilling results (Karig and Ingle et al., 1975). The second reflector (reflector b) is approximately correlated to the boundary between Pliocene and Miocene (Fig. 2). In the Yamato Basin reflector 'b' is lying a little upper horizon from the boundary of the Plio-Miocene. In the northern margin of the Yamato Ridge, reflector 'b' is lying a little lower horizon from the boundary of Plio-Miocene.

The Quaternary layers are composed of clay and diatomaceous clay intercalated by ash and sandy layers in both basins. Contents of sandy layers which consist of turbidites are a little richer in the Japan Basin than in the Yamato Basin. Pliocene sediments are composed of diatomaceous clay in the Japan Basin and clay in the Yamato Basin (Karig and Ingle et al, 1975). Almost the same compositions are commonly observed in the dredged and cored materials in both basins (Koizumi, 1979; Ujiie, 1975; Honza ed., 1979).

The acoustic basement in Yamato Basin is believed to consist of volcanoclastics, judging from the seismic observation. Volcanoclastics of Early to early Middle Miocene ("Green Tuff") are commonly observed at the sedimentary basins in the backarc onshore and continental shelf area of the Tohoku Arc.

The marginal features of the both basins also are different; the sediments of the Japan Basin mostly abut to the east against the foot of the continental

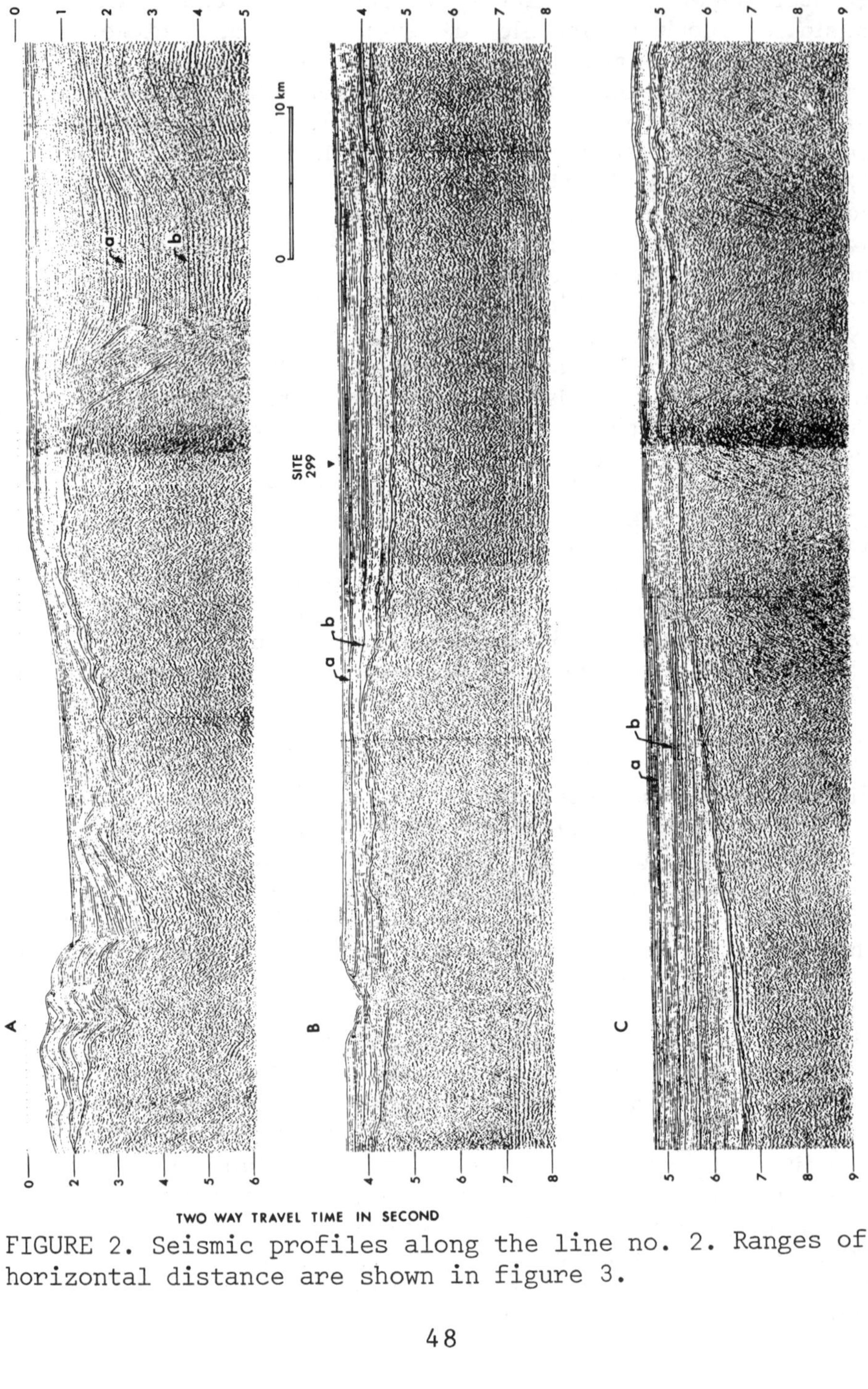

FIGURE 2. Seismic profiles along the line no. 2. Ranges of horizontal distance are shown in figure 3.

slope, whereas the sediments of the Yamato Basin overlap the basement highs at the margins (Fig. 3). The lower transparent layer (Plio-Miocene layer) crops out at the foot of banks and seamounts at the Yamato Basin.

The Tsushima Basin has a smooth floor gently tilting up toward the south. The deepest reflector observed is 2.2 seconds below the sea floor on the foot of the continental slope off San-in. The sedimentary sequence is divided into three units. The upper non-deformed stratified layer with the maximum thickness of 0.3 seconds, the middle stratified and highly reverberant layer with the maximum thickness of 1.4 seconds, and the lower almost evenly deposited weakly stratified layer with more than 1.0 second thick are continuous southwards through the continental slope to the continental shelf off San-in whereas, northwards, they abut onto the basement of the South Korea Plateau. The boundaries between them are slightly disconformable.

The sedimentary sequence with thickness ranging from 0.5 to 1.4 seconds is observed in the Genzan Trough. The northern basement topography is rugged in contrast with the flatness of the southern basement, and the northern basement decreases its depth northward. The northern sedimentary sequence is highly reflective in comparison with the southern sedimentary sequence. These features suggest that the northern part of the Genzan Trough composes the foot of the continental slope and that the terrigenous sediments which are derived from the north form the northern reflective sedimentary sequence. They suggest also that the basement at the foot of the continental slope is dissimilar in origin to the Japan Basin.

The sedimentary sequence of the Tartary Trough is highly reflective in the upper part and its maximum thickness exceeds 2.3 seconds (approximately 2,300m) in the northern part of the trough (Fig. 3). The middle part of the sedimentary sequence of the trough is continuous into the eastern continental slope. The basement is shallower in the southern part of the Tartary Trough, and the depositional features of the sedimentary sequence in this area are irregular.

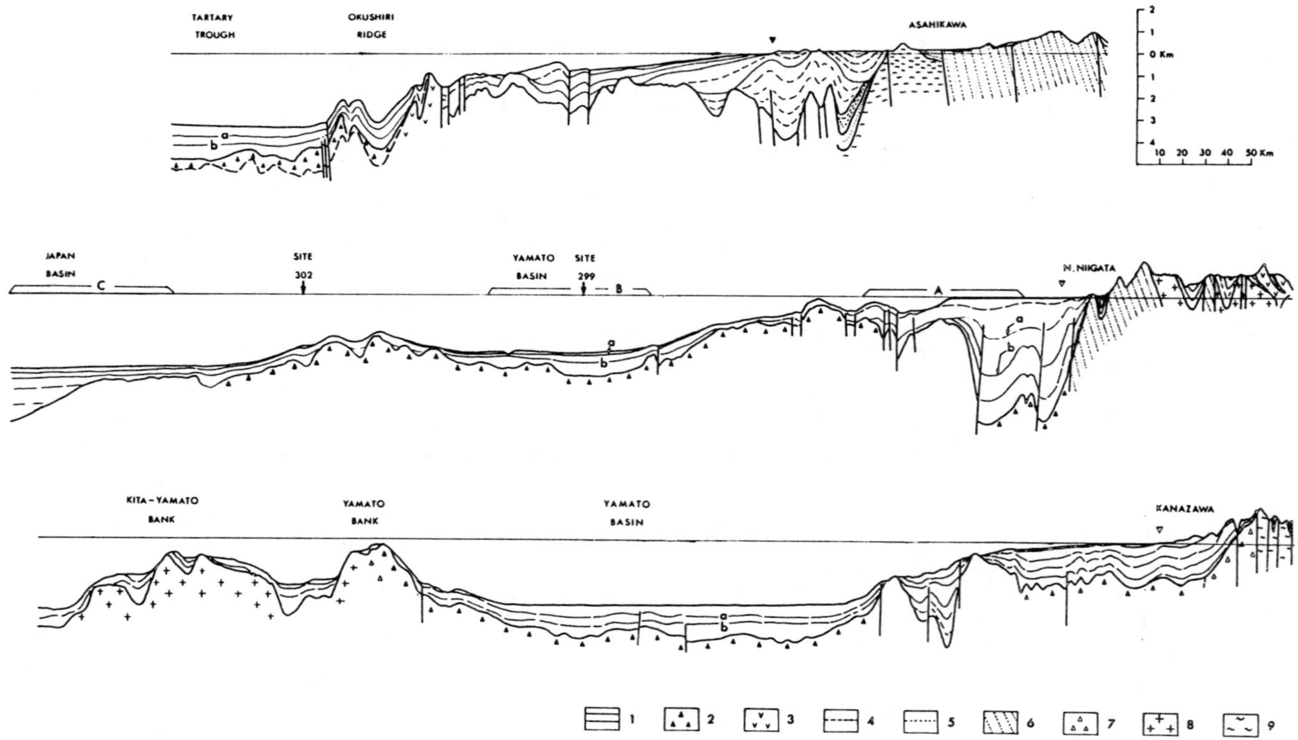

FIGURE 3. Profiles in the Japan Sea (index in Fig. 1). Closed triangles show coasts. Index number 1: Neogene and Quaternary sediments, 2: Neogene volcanoclastics, 3: Neogene and Quaternary volcanic rocks, 4: Paleogene sediments, 5: Cretaceous sediments, 6: Paleozoics, 7: Mesozoic volcanics, 8: granitic rocks, 9: metamorphics.

Two types of acoustic basement are distinguished in the basins of the Japan Sea. One is a weakly stratified layer with a smooth surface, while the other is rugged and may form a lower horizon compared to the former. The acoustic basement, with the smooth surface, has seismic velocities between 2.5-3.5 km/s of CDP stacking velocity, and is widely distributed in the Japan and Yamato Basins. Very little data on the composition of this basement are available. There is a possibility, however as aforementioned, that it may correlate with the Green Tuff of Early to early Middle Miocene age in the Tohoku area. Early Miocene age of the basement appears to be reasonable as compared with the results deduced from the assumption that the sedimentation rates were constant during the Pliocene and Miocene. It should be also noted that there is no identified distribution of marine sediments of Paleogene in the Japan Sea and its surrounding onshore area.

Altered pyroclastic rocks recovered from the deepest horizon at DSDP Site 302 (Karig and Ingle, et al., 1975; Shimazu, 1979) were considered to be recovered from the acoustic basement and once assigned to Green Tuff of Early Miocene, but multi-channel seismic reflection undertaken later indicates that the horizon is considerably above the acoustic basement and near the Plio-Miocene boundary (horizon "b").

Ridges and Plateau

Many topographic highs which characterize the Japan Sea emphasize its complexity compared to other western Pacific backarc basins. Several prominent topographic highs are observed. The Yamato Bank is generally free of sediments but the others are overlain by a sedimentary cover not exceeding 1.0 second.

The Kitayamato and Oki Banks resemble each other in terms of their basement morphology and the acoustic pattern of their sedimentary covers. The acoustic basements of both banks form plateau-like features with rugged summit terraces overlain by densely stratified sedimentary covers. The Kitaoki and Takuyo Banks show a variation in the character of basement highs and depressions (Fig. 3). The basement highs are mostly free of sediments while the basement depressions are filled with over 1.0 second of acoustically stratified sediments which are less stratified than the sediments of the Kitayamato and Oki Banks. Unconformities are observed in the sediments at the Takuyo Bank.

The Yamato Bank forms an ENE trending ridge-like feature with a flat summit, on which the basement is mostly exposed making minor cliffs. This flat summit is interpreted as a wave cut platform formed when the upper parts of the Bank were at sea level.

The Oshima Plateau is different from the banks and ridge mentioned above; the plateau has a slightly rugged summit but with a very thin sedimentary cover. The different character of these topographic highs appears to be related to basement lithologies and their subsidence histories.

The Korea Plateau is covered by sediments of 1.1 seconds in maximum thickness. The sedimentary sequence is divided into three units bounded by disconformities. The middle and lower units are slightly deformed and the upper unit represents sedimentary cover overlying the middle unit. The thickness of the sedimentary sequence is generally thick on the basement depression and on the basement terrace (Tamaki et al., 1978).

Gravels of widely ranging composition (intrusives, volcanics, sedimentary rocks and metamorphic rocks) were dredged from Yamato Ridge (Iwabuchi, 1968; Honza ed., 1979). Some of these are in situ but others may be transported although the area lies at the southern limit of ice-rafting in the Japan Sea. Samples, reflection profiling and magnetic measurements results, all suggest that the Yamato Bank is thinly covered by sedimentary rocks and composed predominantly of granitic rocks with minor intrusives and volcanics. Absolute datings of the granitic rocks give ages of 197 Ma and 220 Ma on Yamato Ridge (Shimazu, 1968), 127 Ma on North Korean Rise (Honza ed, 1978a) and 141.6 Ma on Kitaoki Bank (unpublished data).

Shelves and Slopes along the Japanese Islands

Acoustic basements, folded sedimentary layers, and undeformed sedimentary covers are recognized on the continental shelf and slope. The acoustic basement and the folded sedimentary layers on the continental shelf can be correlated with onshore Neogene volcanics and Neogene sediments respectively, while the undeformed or less deformed sedimentary cover is correlated to Quaternary deposits.

Conspicuous unconformities are observed among these units. The uppermost unit is a well stratified layer which is observed as deposits in the synclinal basins formed by underlying units and in the margin of the continental shelf.

The continental slope to the north of the Noto Peninsula has a complex topography with NNE (south of the Oga Peninsula) to NS trending (north of the Oga Peninsula) ridges and troughs of the Sado Ridge and the Okushiri Ridge. The layers have rather uniform thickness and are deformed and eroded on the Okushiri and Sado Ridges. This clearly suggests that the structural movements which formed these features have been active since the deposition of the upper most layer (Sato, 1971; Hilde and Wageman, 1973; Honza et al., 1977). This movement which is also accompanied by faulting along the continental slope of the Japanese Islands is related to the deepening of Japan Basin. It should be noted that the ridge and trough region is observed only on the eastern margin of the Japan Sea and that NS trending northern region is independent of the trend of the Quaternary Tohoku and Kuril Arcs.

To the west of the Noto Peninsula, the topography is rather simple and quite different from that in the north. Gently folded and densely stratified sedimentary layers of thicknesses less than 1.5 seconds overlie a rugged acoustic basement which, on the upper continental slope, exhibit markedly folded sedimentary layers.

The continental slope area off Hokkaido is complicated by many physiographic features such as the Musashi Bank and the Okushiri Ridge. The acoustic basement on the Musashi Bank may be the product of late Cretaceous volcanic activity (Yuasa et al., 1978). Horst and graben structures of the early Neogene acoustic basement on the slope along the Tartary Trough is analogous to that of the lower continental slope of the Atlantic Ocean (Dingle and Schrutton,

1977), and it is inferred to be foundered continental crust which was rifted during the initiation of oceanic crust within the continent (Montadert, et al., 1977).

SEDIMENTARY BASIN

Many sedimentary basins are observed in the Japan Sea with variable sizes. Some of them are deep sea basins which are presently active in deposition and others are already nonactive in depostion and in some cases destroyed by structural movement. The development of these sedimentary basins is closely related with the tectonics of the Japan Sea since its initiation.

An isopach map of the Japan Sea was compiled from several published and unpublished data (Fig. 4). This map shows the distribution of the sedimentary basins in the Japan Sea excluding margins of Eurasian Continent.

The major deep sea sedimentary basins are the Japan Basin, the Yamato Basin, the Tartary Trough, and the Tsushima Basin, which are presently under active deposition. The thickest sedimentary fill in the Japan Sea, however, is not observed in these deep sea basins but placed at the continental shelf of the Tohoku Arc. The sediment thickness off Niigata exceeds 5.0 seconds and this is the largest one of the sediment thickness of the Japan Sea in our available data.

The sediments are generally thicker to the central part of the Japan Basin and isopach contour of 2.0 seconds shows E-W extension north of the Yamato Bank. Thickest sediment with 2.3 seconds thickness is observed in the center of the basin south of the Bogorov Seamount (Morozowski and Hayes, 1978). The present morphology of the Japan Basin shows an abyssal plain with flat and smooth sea bottom and it indicates that the Basin is presently under the distal turbidites depostion.

The sediment thickness of the Yamato Basin is thinner than that of the Japan Basin. The maximum sediment thickness is 1.6 seconds (Tamaki et. al., 1979), which is about 1.0 second thinner than that of the Japan

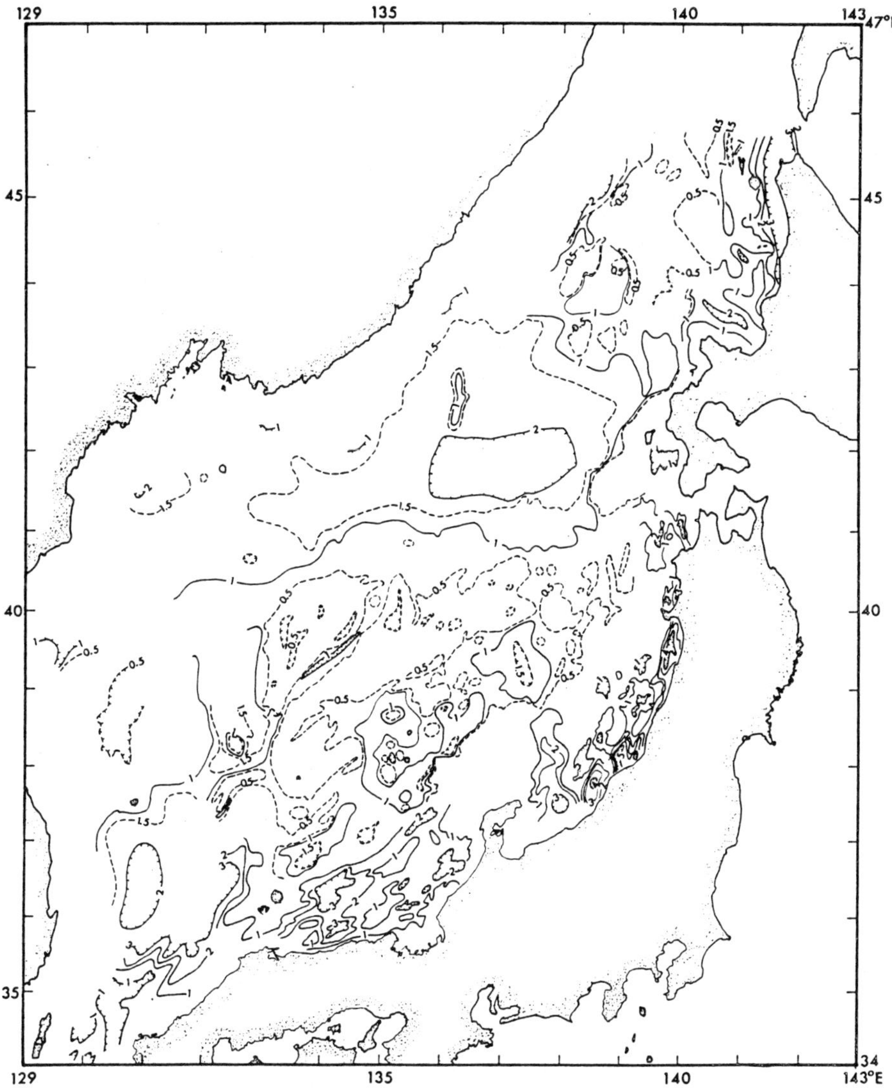

FIGURE 4. Isopach map of the Japan Sea. Contours show sediment isopachous with two-way reflection time in seconds. The map was compiled from Ludwig et al., 1975, Ishiwada and Ogawa, 1976, Morozowski and Hays, 1978, Suzuki, 1979, Tamaki et al., 1979, Tamaki et al., 1981 and unpublished data of the Geological Survey of Japan.

Basin. The difference of the sediment thickness of the both basins suggests different basement ages of them; the younger Yamato Basin and the older Japan Basin. Isopach contours of the Yamato Basin are complicated with many seamounts and subbottom basement reliefs. The sedimentary basin of the Yamato Basin is generally separated into two; the northeastern basin and the southwestern basin. Surface of the present abyssal plain is also separated according to the both sedimentary basins.

The Tsushima Basin at the southwestern end of the Japan Sea has the sediment thickness over than 2.0 seconds. Depositional structure and high reflectivity of the sediments shows that slumping and sliding deposits are main constituents of the sediments in the Basin. Prevailed slumping in the Tsushima Basin may suggest rapid subsidence of the basin.

The sediment thickness of the Tartary Trough exceeds 2.0 seconds. Southern part of the Tartary Trough is dissected by the Tartary canyon. The deposition in the Tartary trough appears to be saturated and the excess sediments are transported to Japan Basin through the Tartary Canyon.

There are many small sedimentary basins observed on the continental shelves along the west coast of the Japanese Islands. The sediment thickness is generally larger than that of the deep sea basin and exceeds 5.0 seconds at maximum as mentioned above. These basins are continuous onto the shore area and make outcrops there. The litho- and biostratigraphy of the basins shows that the sedimentary basins were formed in the deep sea environment since early Middle Miocene.

Judging from paleontological data, invasion of sea water into the Japan Sea area began, at least, in late Early Miocene to early Middle Miocene. Simultaneous occurrence of both deep water and shallow water faunas within a basin (Matoba, 1978; Maiya and Inoue, 1981) shows presence of marked relief on the sea-bottom. And the basins were filled with subtropical or tropical waters which came from the Equatorial Pacific getting over the Japanese Islands.

In the early Middle Miocene the sea increased gradually its depth and the area was followed by a marked change of marine condition. However, most part of the present continental slope of the Tohoku Arc was not below the sea-level until Late Miocene. A marked change of marine condition occurred from early Middle Miocene, namely an influence of boreal waters on both surface and deep water masses which is exemplified by foraminiferal and diatom fossil assemblages. But it is difficult to conclude whether the boreal water came immediately from Pacific side or through opening to the north.

Foraminiferal species charactristic of the Indo-Pacific waters survived to the latest Miocene or early Pliocene, and disappeared when violent compressional tectonic movements occurred widely over the backarc areas, resulting in differentiation of small basins and formation of folds and faults. At the same time as separation from the Pacific deep water masses in the above time, it is inferred from the viewpoint of microbiostratigraphy that bathyal part of the Japan Sea Basins were filled with cold water mass like the present water mass peculiar to the Japan Sea. This means that the present type Japan Sea appeared in Pliocene.

The original depostion of these sedimentary basins in the deep sea environment should be closely related to the formation of the Japan Basin under the tensional tectonics during the early stage of the Japan Sea evolution. The sedimentary basins were uplifted with many thrust faults by E-W compressional tectonics since late Pliocene along the eastern margin of the Japan Sea, and have been separated to many segmented sedimentary basins. Thus, the drastic change of tectonic stress from tensional to compressional occurred at the eastern margin of the Japan Sea in Pliocene. The sedimentary basins on the continental shelf along the eastern margin of the Japan Sea make best oil fields in Japan, the formation of which is closely related to the tectonic history of the Japan Sea (e.g., Ishiwada, 1974; Ikebe et al., 1978).

ARC DEVELOPMENT OF THE JAPANESE ISLANDS

The formation of the Japan Sea might not be an isolated event but an integrated step in the evolution of the Japanese island arcs.

One of the most active arcs during the final development of the Japanese Islands is the Tohoku Arc which has been occurred since the latest Oligocene accompanied by the subduction of the oceanic plate at the front of the Tohoku area (Honza, 1981). The first tectonic stage is followed by volcanic activity along the volcanic chain, erosion in the forearc basin by the uplifted deformation (von Huene, Nasu, et al, 1981) and regional subsidence in the backarc basin. The next stage is subsidence in both of the forearc and the backarc basins during Miocene and Pliocene. The third stage is regional uplift and subsidence to form ridges and troughs in Pleistocene.

The Seinan Japan Arc has been also active since Early Miocene which is suggested in the volcanism along the inner side of the Arc, nevertheless, the subducted slab is not so deep as to reach under the volcanic front (Fitch and Sholtz, 1971) and no active marginal opening is suggested along the inner side of the Arc. The Tsushima Basin is an older one which has not been followed by opening since Pleistocene, at least. However, some regional subsidence might have taken place during Neogene and Quaternary.

Structural movement recognized in the Okinawa Trough in Middle to Late Miocene cannot be seen in the Tsushima Basin, but that in the latest Pliocene to early Pleistocene is recognized on the shelves of San-in. If the both Arcs were connected and were extended as in the same arc system, the difference of the tectonics in the last stage is caused by the difference of subduction condition under the arcs.

The ridge in older stage, the Hidaka Mountains, which is extended throughout the central part of the Hokkaido Island with a N-S trend makes a complexity to understand the Late Mesozoic to Early Cenozoic development of the area. The Hidaka Mountains are suggested to have a paired metamorphic belts of the high temperature and low pressure belt in the east side and of the low temperature and high pressure belt in the west side during Late Mesozoic to Early Cenozoic, and it shows a reversal arrangement as compared with the case in the Seinan Japan Arc (Miyashiro, 1965). The tectonic evolution of Hokkaido should be closely related to the origin of the Japan Sea, although it has been controversial (e.g., Okada, 1983; Kimura et al., 1983).

ORIGIN OF THE JAPAN SEA

An interpretation of the main physiographic elements of the Japan Sea suggests that the ridges and rises correspond to continental structures while the basins represent oceanic features.

The Tsushima Shelf has been under the tectonics of block movement during Neogene where no opening feature is noted. Refraction results show a little different constituents of the basements in both sides along the western depression of the Tsushima Island (Ludwig et al., 1975). It suggests that the tectonic boundary between the older Seinan Japan Arc and the Korea Continent is along the western depression of the Tsushima Islands with NNE-SSW direction, which is a physiographical reflection of thrust faults bounding a half graben to the west (Minami, 1979).

From the paleomagnetic measurements of Cretaceous to Middle Miocene, a clockwise rotation through 58°±14 from 28 Ma to 12 Ma (Otofuji and Matsuda, 1983) or 47° since Cretaceous (Yasukawa, 1975) is suggested in the southwest Japan. A drift of the Tohoku block from south is suggested in the paleomagnetic measurements of Cretaceous rocks (Yasukawa, 1975). The initial position of the southwest Japan is interpreted to be along the Asia continents. For mechanism to drift to south, a fan spreading is proposed in which only the southwest Japan is drifted toward south and the north Japan might be drifted not from the Asia continent but from another place (Honza, 1979).

The initial formation of the backarc basins in the Tohoku Arc is associated with regional subsidence accompanied by normal faultings in Late Oligocene and thick sediments with volcanoclastics are supplied immediately in the backarc side of the volcanic chain. It is remarkable that there is no other place in the Japan Sea where such thick sediments were supplied than the basins immediately backarc side of the volcanic chain, except that in Hokkaido (Figs. 3 and 4). Basins along the center of Hokkaido are interpreted to have been formed as a forearc basin of paleo-Tohoku Arc during Late Paleogene to Early Neogene and then converted to the collision area between the paleo-Tohoku Arc and the paleo-Kuril-Hokkaido Arc (Honza, 1979).

The spreading of the Japan Basin may have occurred in Late Oligocene to Early Miocene, followed by the spreading of the Yamato Basin in Middle Miocene. This feature is suggested in the sedimentary sequence in both basins. All tectonic stages involved a deepening of the Japan and Yamato Basins which are seen both in structural movements and the distributional patterns of the sediments.

The final stage of the Japan Sea is remarkably expressed in the horizontal compressional stress along the eastern margin of the Japan Sea, where several remarkable earthquakes occurred during the past tens of years (Fukao and Furumoto, 1975).

Boundary of the Eurasia and North America plates has been assumed to be through the Arctic Mid-ocean Ridge, Siberia, Tartary Trough, South Hokkaido and to the junction of the Japan and Kuril Trench (Chapman and Solomon, 1976; Dickinson, 1978). A conversion of the Eurasia-North America plate boundary from the initial boundary into the eastern margin of the Japan Sea through Tartary Trough, along the foot of the Tohoku continental slope, Fossa Magna and Suruga Bay is proposed (Kobayashi, in press; Nakamura, in press). The converted age may be concordant with the formation of ridges and troughs along the eastern margin of the Japan Sea. From the seismic profiling results, the ridges and troughs accompanied by thrusts is formed since Quaternary, at least, since late Pliocene. It is inferred that the shortening of the Eurasia and the North America plates occurs along the eastern margin of the Japan Sea with the rate of approximately 1 cm per year in the north Japan (Minster and Jordan, 1978) caused the formation of ridges and troughs accompanied by thrusts since Quaternary and this may imply to create a future trench at the foot of the slope.

REFERENCES CITED

Abe, K. and Kanamori.,1970, Mantle structure beneath the Japan Sea as revealed by surface wave, Bull. Earthq. Res. Inst., v.48, p.1011-1021.

Chapman, M. E. and Solomon S.C., 1976, North-American-Eurasian plate boundary in northeast Asia, Jour. Geophys. Res., v.81, p.921-930.

Dickinson, W. R., 1978, Plate tectonic evolution of North Pacific rim, Jour. Phys. Earth, v.21, p.S1-S9.

Dingle and Schurutton, 1977, Continental margin fault pattern mapped southwest of Ireland, Nature, v.268, p.720-722.

Fitch, T. J. and Scholz, C. H.,1971, Mechanism of underthrusting in southwest Japan: A model of convergent plate interactions, Jour. Geophys. Res., v. 76, p.7260-7292.

Fukao, Y. and Furumoto, M., 1975, Mexhanism of large earthquakes along the eastern margin of the Japan Sea, Tectonopysics, v.26, p.247-266.

Gnibidenko, H., 1979, The tectonics of the Japan Sea, Marine Geol., v.32, p.71-78..

Hilde, T. W. C. and Wageman, J. M.,1973, Structure and oregin of the Japan Sea, Coleman, P. J. ed., The Western Pacific: Island Arc, Marginal Seas, Geo-chemistry, Univ. W. Australia Press, p.415-434.

Honza, E., 1979, Distrtibution and characteristic features of the sedimentary basins in the island arc, Jour. Japan. Assoc. Petrol. Tech., v.44, p.352-361 (in Japanese with English abstract).
Honza, E., 1981, Subduction and accretion in the Japan Trench, Oceanologica Acta., Sp. no., p.251-258

Honza, E., ed., 1979, Geological investigation of the Japan Sea, Cruise Rept., no.13, 99p Geol. Surv. Japan,.

Honza, E., Kagami, H. and Nasu, N., 1977, Neogene

geological history of the Tohoku (NE Japan) Arc, Oceanog, Soc. Japan, V.33, p.297-310.

Ikebe, Y., Katahira, T. and Miyazaki, H., 1978, Some problems on petroleum geology in Japan, in Fujita, K. et al., eds.,Cenozolc Geology of Japan, p.205-216.

Isezaki, N., 1976, Possible spreading center in the Japan Sea, Jour. Marine Geophys. Res., v.4, p.53-65.

Ishihara T., in press, Gravity field around Japan - sea gravimetry by the geological Survey of Japan, Mar. Geod.

Ishiwada, Y., 1974, Petroleum geology of the continental margin adjascent to Japanese Island, Jour. Japan. Assoc. Petrol. Tech., v.39, p.216-220.

Ishiwada, Y. and Ogawa, K., 1976, Petroleum geology of offshore areas around the Japanese Islands, CCOP Technical Bulletin, v.10, p.23-34.

Iwabuchi, Y., 1968, Submarine geology of the southeastern part of the Japan Sea, Tohoku Univ. Inst. Geology and Paleontology Contr., no.66, p.1-76 (in Japanesewith English abstract).

Karig, D.E. and Ingle Jr., J.C. ed., 1975, Initial reports of the Deep Sea Drilling Project, v.31, U.S. Govt. Printing Office, Washington D.C., 927pp.

Karig, D.E., 1973, Comparison of island arc-marginal basin complexes in the north-west and south-west Pacific, in Coleman, P. J. ed., The Western Pacific: Island Arc, Marginal Seas, Geo-chemistry, Univ. W. Australia Press, p.355-364.

Kimura, G., Miyashita, S., and Miyasaka, S., 1983, Collision tectonics in Hokkaido and Sakhalin, in M. Hashimoto and S. Uyeda, eds., Accretion Tectonics in the Circum-Pacific Regions, Terra Sci. Publishing Co., p.123-134.

Kobayashi, Y., in press, Conditions for the beginning of subduction, Chikyu (in Japanese).

Koizumi, I., 1978, Neogene diatoms from the Sea of Japan, Marine Geol., v.26, p.231-248

Ludwig, W.J., Murauchi, S., and Houtz, R.E., 1975,

Sediments and structure of the Japan Sea, Geol. Soc. Amer. Bull., v.86, p.651-664.

Maiya, S. and Inoue, Y., 1981, Fossil foraminiferal assemblages from the Middle Miocene of the Niigata sedimentary basin and evolutional change of paleogeography, Rept. Tech. Lab. JAPEX, v.24, no.2, p.139-143 (in Japanese).

Matoba, Y., 1978, Evolutional change of paleoenvironment of the Japan Sea from a viewpoint of benthic and planktonic foraminifers, Kaiyo Kagaku (Marine Sciences), v.10, no.42, p.269-277, (in Japanese).

Minstar, J.B. and Jordan, T.H., 1978, Present-day plate motions, Jour. Geophys. Res., v.83, p.5331-5354.

Miyashiro, A., 1965, Metamorphic rocks and metamorphic belts, Iwanami Book Co., Tokyo.(in Japanese)

Montadert, L., et al., 1977, Rifting and subsidence on passive continental margins in the North East Atlantic, Nature, v.268, p.305-309.

Mrozowski, C.L. and Hayes, E.H., 1978, Sediment isopachs, in Hayes., E., ed, A Geophysical Atlas of the East and Southeast Asian Seas, MC-25, Geol. Soc. Amer. .

Murauchi, S., 1972, Crustal structure of the Japan Sea measured by refraction method, Kagaku, v.42, p.367-365 in Japanese).

Nakamura, K., in press, Possible recent trench along the eastern Japan Sea as the convergent boundary between the Eurasian and American plates, Bull. Earthq. Res. Inst., Tokyo Univ.

Okada, H., 1983, Collision orogenesis and sedimentation of Hokkaido, Japan, in M. Hashimoto and S. Uyeda, eds., Accretion Tectonics in the Circum-Pacific Regions, Terra Sci. Publishing Co., p.91-106.

Otofuji, Y. and Matsuda, T., 1983, Paleomagnetic evidence for the clockwise rotation of southwest Japan, Earth Planet. Sci. Letters, v.62, p.349-359.

Sato, T., 1971, Sea bottom survey in westward of the

Northeast Japan, Jour. Geogr. Soc. Japan, v.80, p.285-301.

Shimazu, M., 1968, Absolute age determination of granite in Yamato Bank, Sci. Letters on Res. Japan Sea, no.6, p.55-56 (in Japanese).

Shimazu, M., 1979, "Green tuff" in the core from DSDP Leg 31 Site 302: Kita-Yamato tai, Jour. Geol. Soc. Japan, v.85, p.655-656 (in Japanese).

Soloviev, S.L., Touzov, I.K., and Vasiliev, B.I., 1977, The structure and origin of the Okhotsk and Japan Sea abyssal depressions according to new geophysical and geological data, Tectonophysics, v.37, p.153-166.

Suzuki, U., 1979, Petroleum geology of the Sea of Japan, northern Honshu, Jour. Japan. Assoc. Petrol. Tech., v.44, p.292-307 (in japanese with English abstract)

Tamaki, K., Honza, E., Yuasa, M., Nishimura, K., and Murakami, F., 1981, Geological map of the central Japan Sea, Marine Geol. Map Ser., no.15, Geol. Surv. Japan.

Tamaki K., Murakami, F., Nishimura, K., and Honza, E., 1979, Continuous seismic reflection profiling survey, in Honza, E., ed., Geological Investigation of the Japan Sea, Cruise Rept., no.13, p.48-51,Geol. Surv. Japan.

Tamaki, K., Yuasa, M., Nishimura, K., and Honza, E., 1979, Geological map of the Japan and Okhotsk Seas around Hokkido, Marine Geol. Map Ser. no.14, Geol. Surv. Japan.

Tomoda, Y., 1973, Maps of free air and Bouguer gravity anomalies in and around Japan, Tokyo Univ. Press.

Tuezov, I.K., 1971, Crustal structure of the Okhotsk and Japanese area from regional seismic prospecting data, Asano, S. and Udintsev, G.B., eds., Island Arc and Margianl Sea., Tokai Univ. Press, p.121-135.

Ujiie, H., 1975, Geological history of the Japan Sea, revealed by sediments. Nat. Sci. Mus., no.42, p.55-66 (in japanese).

von Huene and Nasu, N., 1978, Japan Trench transected, on Leg 57, Geotimes, v.23, no.4, p.16-21.

Yasukawa, K., 1975, Paleolatitude and relative position of South-West Japan and Korea in the Cretaceous, Geophys. Jour. Royal Astr. Soc., v.43, p.835-846.

Yoshii, T, 1972, Features of the upper mantle around Japan as inferred from gravity anomalies, Jour. Physics Earth, v.20, p.23-24.

Yuasa, M., Tamaki, K., Nishimura, K., and Honza, E., 1978, Welded tuff dredged from Musashi Bank, northern Japan Sea and its K-Ar age, Jour. Geol. Soc. Japan, v.84, p.375-377.

STRUCTURE AND TECTONO-MAGMATIC EVOLUTION OF THE PHILIPPINE SEA FLOOR

SHARASKIN, A.YA.
V.I.Vernadsky Institute of Geochemistry and Analytical Chemistry, USSR Academy of Sciences, Moscow, USSR.

INTRODUCTION

The Philippine Sea is a significant link of the active margin system of the Western Pacific. The first geological review of the region, attempted by H.Hess (17), revealed the strikingly regular arrangement of the structural elements of the sea-floor, which attracted keen attention. During last two decades the region has been studied by a wide number of soviet, japanese and american marine expeditions, including five DSDP Legs. The latters provided the most valuable and informative data, discussed in initial Reports of DSDP (14,21,26,29,30), and in some other publications (5,15,8, 47).

The main point under discussion was a problem of the agreement between the observed facts and the theoretical model of the subduction derived geodynamics. Indeed, all recent achievements in geosciences left no doubts about the plate convergence all along the active margins, however the existing ideas on what is happening down below are still rather dim, even when the main principles are under consideration.

This paper presents a synthesis of available data, elucidating the crustal and lithospheric parameters of the Philippine Sea region, the major stages in its geological evolution and geochemistry of volcanic rock series, erupted in the course of the latter. Some geodynamic problems, arising from new data set will also briefly be discussed.

Proceedings of the 27th International
Geological Congress, Volume 23, pp. 67–92
ORIGIN AND HISTORY OF MARGINAL AND INLAND SEAS

CRUSTAL STRUCTURE AND COMPOSITION

The Philippine Sea floor (Fig.1), surrounded by paired systems of trenches and active island arcs, consists of several basins separated by narrow, alongated ridges. The average depth of the basins is regularly decreasing from the west to the east, and crest lines of ridges as well become shallower in this direction (2,33).

One of the significant structures is the Palau-Kuyushu Ridge, subdividing the region into two parts with the different orientation of the major relief forms. The latter is not accidental but coincides with the lineations of the magnetic anomalies west and east off the ridge. In the West Phillipine Basin the observed magnetic anomalies are symmetrical relative to the Central Basin Fault, and in the Shikoku and Parece Vela Basins they have the Kinan Seamounts and IPOD Fracture Zone as their axes of symmetry (31, 36, 58), thus enabling an interpretation of these structures as extinct spreading centres.

The suggested structural subdivision of the area is further supported by seismological and gravimetrical data. The crustal sections, constructed on a base of the seismic refraction studies (Fig.2), illustrate perfectly, that in all basins the crust is definitely of oceanic type. However, its thickness almost never exceeds 6 km, consequently being much thinner than an average about 7.5 km thick crust of the North-Western Pacific (37). The crustal thickness increases up to 9-15 km under the ridges and rises, mainly due to thickening of the seismic layer 2, which has in some cases also more complex seismic parameters, while the layer 3 remains rather uniform both in thickness and in distribution of seismic velocities. Seismologists assume the presence of thin isolated lenses with characteristics of "granitic layer" in crustal sections under the Amami Plateau, Daito Ridges, Northern Iwo-Iima Ridge and beneath the northern and southern parts of the Kuyusku-Palau Ridge. In all other areas only oceanic layers 2 and 3 have been recognized (32,37).

According to seismoprofiling data, discussed by some authors (e.g., 25), the sedimentary cover in the basins of the Phillipine Sea is generally thin (110-200 m), but near to the western slopes of the Kuyushu-Palau, West Mariana and Mariana Ridges the thickness of sediments regularly increases, being in certain cases more than 1 km. The last feature, interpreted as a result of the accumulation of the volcanoclastic material, is now directly evident by DSDP results. Interestingly, these volcanogenic aprons are present only west but not east off ridges, where only pelagic sediments overlay the acoustic basement. No folding deformations have been observed in the basin sediments. In trenches the sediment distribution is also assymmetrical: the outer trench slopes are covered by normally bedding oceanic sediments, while spotty deposits are characteristic for inner slopes, including besides others turbidites and talus breccias. There are no thick accretionary prisms of sedimentary material neither at the base of inner slopes nor at the trench bottom (21, 22).

The numerous rock samples, recovered from various structures of the Philippine Sea (see Fig.3 for locations of sampled areas), indicate that the crust of basins is different in composition from that of ridges. The DSDP results proved that the acoustic basement of the basins consists of basaltic lavas, which are most probably overlaying, like in oceanic regions, the diabase - gabbro - serpentinite complexes. In the West Philippine Basin the latter rock types were encountered at the DSDP site 293, located in the Central Basin Fault zone (26), and, in the Parece Vela Basin they were dredged from canyon walls of the IPOD Fracture Zone (40). These facts again support the above idea, that both zones are analogues of spreading centres.

The rocks, recovered from ridges and rises are much more valiable in composition, Roughly they group into four assemblages:

1. The ophiolitic rocks, dredged from Amami Plateau, Daito Ridges and from inner walls of Yap and Mariana Trenches: most probably these rocks are forming the Mesozoic basement of the structures under consideration (15,40,52,54).
2. Plagiogranites and acid metamorphic rocks, dredged from the above areas and from the Kuyushu-Palau Ridge as well. The K-Ar dating indicate their age interval in between

150-15 Ma (11-13, 15,35,40,52,54), and this scattering results both from repeated metamorphism and from different structural positions of the rocks: some of them came from the basement complexes, and others reflect the later intrusive activity.

3.Cenozoic volcanics of island are affinity, recovered everywhere on ridges and piling up their upper crustal levels (51).

4.Lithified sedimentary rocks of Cenozoic age, tightly associated with the previous rock group.

Hence it appears clear, that now inactive ridges were volcanically active structures in the past, similar in many respects to the current island arcs. The alternate arrangement of these structures and basins floored by oceanic crust suggests the rather complex geological evolution of the region and the remittent character of the related magmatism.

THE UPPER MANTLE STRUCTURE

Inspite of the evident crustal heterogeneity, the Philippine Sea region reacts as a single lithospheric plate, rimed by narrow belts of high seismicity and volcanic activity, when its inner areas are practically aseismic and have no active volcanoes. The spatial configuration of the Benioff zones is complicated (27): the eastern one is relatively gently dipping in the north, but becomes nearly vertical below 200 km in the south. The deep-focus earthquakes occur here down to 16°N, and further to the south seismicity and epicenters depths are sharply decreasing. The western zone is definitely segmented; the deep shocks are observed only in its southern part, and north off Mindanao Island, they are not deeper than 280 km. The distance between seismic planes and the feet of volcanoes ranges non-systematically from 100-200 km. The frequency of events is also variable both along and across zones.

The heat flow measurements (3,36,43,55,56) indicate the active thermal state of the earth's interior beneath the vast portion of the Philippine Sea floor (Fig.3). Due to high scatter of measured values, the mean heat flow in the region is difficult to estimate, but clearly it has to be higher than that predicted by a simple cooling of the lithosperic plates (39,43,61), leading to the suggestion on the decreased thickness of the Philippine Sea lithosphere and uprise of hot mantle material towards its base. This statement is directly supported by electrical conductivity anomalies, observed beneath the northern Shikoku Basin (19,41), and by analyses of the group velocities of Love and Rayleigh waves, induced by large earthquakes (1,45). According to these studies the lithosphere thickness in the region is about 30 km, i.e. it is at least twice thinner than the Pacific plate nearby, and the underlying mantle has abnormally low velocity characteristics (45).

The gravimetrical observations (57) detect the intensive disturbance of gravity field all along the boundaries of the Philippine Sea, and suggest a small average value of the positive isostatic anomaly for its internal part, again evidencing the assumed uprise of hot mantle underneath.

Summarizing the above geophysical evidences, one may conclude, that abnormally thin lithispheric plate of the Philippine Sea, covered by the reduced oceanic crust, undergoes the intensive transformation along its outer zones, interacting with island arc systems on the west and with thick oceanic plate on the east. The absence of any significant deformations in its inner areas can be explained in no way besides the extentional regime, prevailing in the course of the plate evolution due to certain kind of endogenous geodynamic factors.

MODEL OF THE GEOLOGICAL EVOLUTION

Such as above anomalous crustal and lithosperic parameters of the Philippine Sea plate are postulated as resulting from a peculiar geological evolution of the entire area. Unfortunately, there are many unsolved problems pertinent to the origin of the biggest West Philippine Basin (20), but obviously it is not the entrapped fragment of a normal oceanic plate, as was firstly suggested (6). Recent interpretations of the magnito-metric data (31) and DSDP results imply like in other cases this basin's crust being newly formed behind the island arc.

The idea on marginal basin genesis due to back-arc seafloor spreading belong to Karig (23,24). Developing it on a base of the first drilling results in the Philippine Sea (25) he pointed out the following important regularities:

1) the younger the age of the acoustic basement, the more remote is the basin from the continent, that suggests a periodical displacement of the back-arc spreading zones;
2) the ridges, separating the basins, are formerly active island arcs, now having volcanoclastic aprons only on one side, faced toward the older basin, that implies a repeated splitting of the arc structures;
3) geology of the Mariana Islands indicates an undulating arc volcanic activity with three main pulses in due Cenozoic time.

On this basis Karig presumed that new basins began to form after the rifting of an arc, followed by its frontal segment displacement toward the ocean, when the rear segment remained as a bordering structure of the former marginal Sea, and an opening space was flooring by younger oceanic crust. Accordingly the arising structures were termed remnant arcs, inter-arc basins, back-arc spreading centers etc. One of the important points in Karig's speculations was the notion on time coordination between arc volcanic pulses and spreading activity behind the arc, that is why both events were announced as coming out the increase of the subduction rate.

This simple and logical model was proved in many principles by recent much more detailed data. The only serious modification is concerning the problem of time relations between periods of the arc volcanism and back-arc spreading. According to new interpretations (see explanations to Fig.1 for inferred time intervals) these events are not synchronous, but rather have taking turns one after another. It is worth to note as well that the arc thrusting over the oceanic plate is evidenced now as non-forming the accretionary prisms fo sediments (20). In contrast, the frontal arc zones are likely to be intensively destructed, and the disintegrated material is sinking underneath the arc base together with the oceanic plate.

GEOCHEMISTRY OF THE MAIN VOLCANIC ROCK SERIES

The current views on the genesis of the Philippine Sea floor suggests the very important role of magmatism, strictly forming the crust of the basins and upper structural levels of the island arcs. As it was already pointed out, the crustal layer 2 in all basins consists of basalts and recent comprehensive geochemical studies (e.g., 7,16) revealed their strong similarity to mid-ocean ridge basalts (MORB), thus perfectly supporting the idea on the basin genesis due to sea-floor spreading. The only point, remaining under discussion, is the character of distribution of the trace elements like K, Rb, Ba and Sr. Indeed, in a number of cases there are high concentrations of these elements in the inter-arc basalts, and some researchers suggested the primary origin of such geochemical anomalies subsequently implying a peculiar geodynamics beneath the back-arc spreading center (34,38). However, exactly the above trace elements could be easily sorbed by rocks from the seawater even due to very slight halmyrolisis or hydrothermal alterations, and the reported anomalies should be considered first of all on this assumption. Otherwise one can hardly explain why having the deep-seated source of the anomalies the same spreading center has erupted both the normal and modified magmas.

The geochemical analogy of MORB and inter-arc basin basalts in many highly important parameters is illustrated on Figures 5-8. The presented diagrams display as well the very distinctive geochemistry of the near-by arc volcanics. The selected examples like other available data indicate the genetic diversity of all arc volcanic series, observed in the region. For example, the Ta/Th ratios in calc-alkaline volcanics are systematically lower than in the rocks of the arc-tholeiite series, and both rock types are different in this respect from MORB or inter-arc tholeiites (Fig.6). At the same time within each series under comparison those ratios remain relatively constant in spite of all peculiarities of magma differentiation processes and possible secondary alterations, thus excluding anything but the primary nuances in magma compositions.

This conclusion is further evidenced by the fine distinction in the relative abundances of other trace elements like K,Rb,Ba,Sr, REE, Zr, Hf etc. (Figures 5 and 7). In this respect, it is of importance to note that all the arc volcanics, having prominent enrichment in some magmaphil elements, are depleted in the others (Fig.7) in spite of the usually coherent behaviour of all these elements in the magmatic processes. Boninites, recently attracted special attention, having some more inherent features. Chemistry and mineralogy of these rocks evidence the origin of this magma types from the ultramafic mantle material under the high water pressure (e.g., 48), that is why these olivine and pyroxene phyric lavas have at the same time the glassy matrix abnormally rich in silica and water. The point of interest is that correlated $^{143}Nd/^{144}Nd$ and $^{87}Sr/^{86}Sr$ data (Fig.8) argue against the seawater as affecting the parent magma composition.

SOME PROBLEMS OF THE ACTIVE MARGIN GEODYNAMICS

All recent advances in marine geology imply with no doubt the plate convergence along the active margin systems. In respect to geodynamics of these zones, this makes many researchers, including those who studied the Philippine Sea, to believe that the active subduction of the oceanic plates is the only factor, responsible for the complex geological evolution of the areas under consideration. However, trying to defend the unambiguous significance of subduction one has to solve a number of serious problems, which is attainable only by means of very complex adaptations of the originally simple and logical model.

First of all, the subduction, being claimed as the necessary process to compensate the sea-floor spreading along the mid-ocean ridges, at the same time has to be considered as putting into operation the new spreading zones behind the island arcs. The question to the point is why the latter zones should be secondary without fail? The summary input of the back-arc spreading activity into crustal growth is not al all neglected, and the Philippine Sea is a good example of it. The total area of the newly formed oceanic crust in this region is quite comparable to the productivity of the mid-ocean ridge zone of the same length in due equal time intervals.

The original model of subduction is also discording with the observed periodicity of both tectonics and magmatism of the active margins. The only way to avoid that obstacle is an assumption on the episodic increase of the subduction rate as only then activating the pertinent processes (e.g., 22,25). In this case the induced back-arc spreading would speed up the subduction further more in consequence of the required arc thrusting over the oceanic plate, and the whole progression of the events should become self-accelerating. But in fact it is rather slowing down, at least in the Philippine Sea, and one has to introduce the additional contrivances in order to get the desirable result. The suggested effect of the bathymetric rises as ceasing the subduction (28) doesn't help very much, since it is neither counting the plate inertia nor explaining the regularity of the events, and moreover it parallels to the widespread notion that even mid-ocean ridges may sink beneath the overriding plate.

There is a number of theoretical considerations of the processes, caused by the cool rigid lithosphere descending into the hot plastic mantle (e.g.53). They show that the sinking plate would be principally able to provoke the secondary mantle deapirism, but the final results in such modeling are highly dependent on the initially selected parameters, and the problem of systematic alternation of the geological events in the active margin evolution still remains unbeaten.

With the above problems in hands many workers consider the geochemical data as the most significant indicators of the subduction. The idea is based on strong geochemical differences between subducted crustal materials (including alteration products, sediments, and seawater) and the upper mantle, therefore magmas generated within or above the subduction zones have to be unlike the purely mantle derivates. (42).

Data of the previous section and many published results (7,16,46,47,59) evidence the lack of geochemical anomalies definitely pointing to the influence of the subducted materials of any kind on the composition of the back-arc basin basalts. Those, which have been observed, are doubtful, and the subduction induced origin of the back-arc spreading centers is left under suspicions.

Geochemistry of the island arc volcanic series put into straits being quite contradictive. On the one hand, all these series are definitely enriched in certain components which could be suspected as being introduced down to the magma sources due to subduction. These arc volatiles, including H_2O, and large ions lithophil elements like K,Rb, Sr, and Ba. On the other hand, another magmaphil element with low partition coefficients between crustal phases and melts surprisingly does not follow the same course. As it was shown earlier, the island arc volcanics are often abnormally depleted in Zr, Hf, Nb and Ta, and some times in REE as well. The latter anomalies totally exclude the variant of the possible remelting of the subducted crust or sediments, and thereby forces the advocates of the viewpoint under discussion to make provision for these elements cutting off island arc magmas. Due ideas do not go further than the supposed stabilization of the suitable accessory minerals in the mantle because of the water pressure increase on account of subduction (38). The idea is not seriously tested, and therewith puts forward the new problem of the alien component accumulation in the interiors of the active margins.

For reasons given above the widely accepted meaning of subduction seems to be quite exaggerated and needs a revision. Admitting the basic principle of the plate convergence, one can avoid the above difficulties by means of two simple assumptions. Firstly, it is necessary to give certain credit for the endogenous mantle activity non-produced by, but definitely interacting with the subduction; only this way makes it easier to understand the very active geodynamic state of the active margin zones. And secondly, one ought to regard only the lower part of of the oceanic lithosphere as sinking deep into the mantle while its crustal portion is to be piled-up tectonically beneath the leading edge of the overriding plate; only then it will be possible to discount all geochemical contradictions.

REFERENCES

1. Abe K. and Kanamori,H. 1970. Upper mantle structure of the Philippine Sea. In: Island Arc and Ocean. A symposium. M.Heshino and H.Aoki Eds., Tokai University Press: 85-92.
2. Agapova G.V., Beljaev A.V., Perevozchikov A.V. et al. 1980. Geomorphology of the floor and structure of the 1st layer of oceanic crust. In: Geology of the Philippine Sea Floor, A.V.Peive (Ed.), Nauka, Moscow: 13-37 (in Russian).
3. Anderson R.A., Langseth M.G., Hayes D.E. et al. 1978. Heat flow, thermal conductivity, thermal gradient. In: Geophysical Atlas of East and Southeast Asian Seas, Hayes D.E., Ed., Geol.Soc.Am. Map and Chart Series, MC-25.
4. Aoki H. and Iwabuchi Y. 1972. Submarine geology of northern region of Izu-Bonin Ridge. In: Izu Peninsula. M.Hoshino and H.Acki (Eds.) Tokai Univ. Press:311-319.
5. Bogdanov N.A. and Khain V.E. 1980. General remarks on the earth crust evolution of the Philippine region. In: Geology of the Philippine Sea Floor, A.V.Peive, (Ed.), Nauka, Moscow: 243-250 (in Russian).
6. Ben-Avraham Z., Bowin C, and Segawa J. 1972. An extinct spreading centre on the Philippine Sea, Nature, 240, 453-455.
7. Bougault H, Maury R.C. El Azzouzi M et al. 1981. Tholeiites, basaltic andesites, and andesites from Leg 60 sites: Geochemistry, mineralogy and low partition coefficient elements. In:Hussonf D.M., Uyeda S. et al. Init. Repts. DSDP, 60, Washington (US Gov. Printing Office): 657-677.
8. Crawford A.J., Beccaluva L. and Seri G. 1981. Tecto-

no-magmatic evolution of the West Philippine-Mariana region and the origin of boninites. Earth Planet. Sci. Lett., 54:346-356.

9. De Paolo D.J, and Wasserburg G.J. 1977. The sources of island arcs as indicated by Nd and Sr isotopic studies Geophys.Res. Lett., 4, 465-468.

10. Dixon T.H. and Batiza R. 1979. Petrology and chemistry of recent lavas in the Northern Marianas: Implications for the origin of island arc basalts. Contrib. Mineral. Petrol., 70, 167-181.

11. Evlanov Yu.B., Konovalov Yu.I., Svinnikov A.I. 1979. On geological structure of some rises on the Philippine Sea floor. In: Geology of the Sea of Japan and the Philippine Sea. Vladivostok, 21-35 (in Russian).

12. Evlanov Yu.B., Lipkin Yu.G., Azmuk S.V. et al. 1977. Some features of geology of the Sea of Japan and the Philippine Sea. In: Geological studies in marginal seas of N.W.-Pacific. Vladivostok, 31-39 (in Russian).

13. Evlanov Yu.B., Mishkin M.A., Tararin I.A. 1978. Metamorphic complex of the southern Kyushyu-Palau Ridge (the Philippine Sea). In: Geology of the Sea of Japan and the Philippine Sea. Vladivostok, 26-36 (in Russian).

14. Fischer A.G., Heezen B.C et al. 1971. Init. Repts. DSDP, 6, Washington (US Govern. Printing Office).

15. Gnibidenko H.S. 1979. The tectonics of the Far East marginal seas. Nauka, Moscow, p.161.

16. Hawkins J.W. 1977. Petrologic and chemical characteristics of marginal basin basalts. In Talwani M. and Pitman W.C.III (Eds.), Island Arcs, Deep Sea Trenches and Back-Arc Basins, Washington (Am. Geophys. Union), p. 355-365.

17. Hess H.H. 1948. Major structural features of the western North Pacific: an interpretation of H.0.5485, bathymetric chart, Korea to New Guinea. Geol.Sco. Am. Bull., 59, 417-466.

18. Hichey R.L. and Frey P.A. 1981. Rare-earth element geochemistry of Mariana fore-arc volcanics: DSDP Site 458 and Hole 459. Init. Repts. DSDP, v.60. Washington, (US Goverment Printing Office), 735-742.

19. Honkura Y. 1974. Electrical conductivity anomalies

beneath the Japan Arc., J.Geomag.Geoelectr., 26, 147-171.
20. Hussong D.M. and Uyeda S. 1981. Tectonic processes and the history of the Mariana Arc: A synthesis of the results of Deep Sea Drilling Project Leg 60. Init.Repts. DSDP, 60, Washington (US Gov. Printing Office), 909-929.
21. Hussong D.M., Uyeda S. et al. 1981. Init.Repts DSDP, 60, Washington (US Gov. Printing Office).
22. Jurdy D.M. 1979. Relative plate motions and the formation of marginal basins. J.Geophys. Res., 84, 6796-6801.
23. Karig D.E. 1970. Ridges and basins of the Tonga-Kermadec island arc system. J.Geophys. Res., 75, 239-254.
24. Karig D.E. 1971. Structural history of the Mariana island arc system. Geol.Soc. Am.Bull. 83, 323-344.
25. Karig D.E. 1975. Basin genesis in the Philippine Sea. In: Ingle J.C., jr. , Karig D.E. et al. Init. Repts, DSDP 31: Washington (US Gov. Printing Office), 857-879.
26. Karig D.E.,Ingle J.C.,jr. et al. 1975. Init.Rept. DSDP 31: Washington (US Gov. Printing Office).
27. Katsumata M. and Sykes L.R. 1969. Seismicity and tectonics of the Western Pacific: Izu-Mariana-Caroline and Ryukyu-Taiwan regions. J.Geophys. Res., 74, 5923-5948.
28. Kelleher J. and McCann W. 1977. Bathymetric highs and development of convergent plate boundaries. In: Island arcs: Deep Sea Trenches and Back-Arc Basins. M. Talwani and W.C.Pittman (Eds.). M.Ewing Ser.1, Washington (AGU), 115-122.
29. Klein G., Kobayashi K. et al. 1980. Init Repts.DSDP 58, Washinton (US Gov.Printing Office).
30. Kroenke L., Scott R. et al. 1980. Init. Repts. DSDP 59, Washington, (US Gov. Printing Office).
31. Louden K.E. 1976. Magnetic anomalies in the West Philippine Basin. In: Sutton G.H., Manghnami M.H. and Moberly R, (Eds.). The Geophysics of the Pacific Ocean Basin and its margins. Geophys. Monogr. Ser.19, Washington (AGU), 253-267.
32. Ludwig W.J., Murauchi S., Den N. et al. 1973. Structure of East China Sea - West Philippine Sea margin

off Southern Kuyshu, Japan. J.Geophys.Res. 78, 2526-2536.

33. Mammericks J, Fisher R.L., Emmel P.J. et al. 1977. Bathymetry of the East and Southeast Asian Seas. Geol. Soc.Am. Map. Chart Series, MC-17.
34. Mathey D.P., March N.G. and Tarney J. 1980. The geochemistry, mineralogy and petrology of basalts from the West Philippine and Parece Vela Basins and from the Palau-Kyushu and West Mariana Ridges. In: Kreenke, L., Scott R. et al., Init.Repts. DSDP, 59, Washington (US Gov. Printing Office), 753-800.
35. Mizuno A, Okuda Y, Tamaki K. et al. 1975, Marine geology and geologic history of the Daito Ridges area, Northwestern Philippine Sea. Marine Sci. Monthly, 7, N 7 and 8, 52-58 and 39-44.
36. Mrozowski C.L., Hayes D.E. 1979. The evolution of the Parace Vela Basin, eastern Philippine Sea. Earth Planet. Sci. Lett., 46, 49-67.
37. Murauchi S, Den N., Asano S. et al. 1960. Crustal structure of the Philippine Sea. J.Geophys. Res., 73, 3143-3171.
38. Natland J.H. and Tarney J. 1981. Petrologic evolution of the Mariana arc and back-arc basin system.- A synthesis of drilling results in the South Philippine Sea. Init.Repts. DSDP 60, Washington (US Gov. Printing Office), 877-908.
39. Parsons B. and Sclater J.G. 1977. An analysis of the variation of ocean-floor bathymetry and heat flow with age. J.Geophys. Res., 82, 803-827.
40. Peive A.V.(Ed.) 1980. Geology of the Philippine Sea floor. Nauka, Moscow, pp 259. (in Russian).
41. Rikitake T. 1969. The undulation of an electrically conductive layer beneath the islands of Japan. Tectonophysics, 7, 257-264.
42. Ringwood A.E. 1975. Composition and petrology of the Earth's mantle. McGraw-Hill, Inc., New York, p.322.
43. Sclater J.G., Karig D., Lawyer L.A. et al. 1976. Heat flow, depth and crustal thickness of the marginal basins of the South Philippine Sea. J.Geophys. Res., 81, 309-318.
44. Scott R.E., Kroenke L., Zakariadze G. et al. 1980. Evolution of the South Philippine Sea: DSDP Leg 59

results. In: Kroenke L, Scott R.B. et al. Init. Repts. DSDP, 59, Washington (US Gov. Printing Office), 803-816.
45. Seekins L.C. and Teng T.L. 1977. Lateral variations in the structure of the Philippine Sea plate. J.Geophys. Res., 82, 317-324.
46. Sharakin A.Ya. 1981. Petrography and geochemistry of basement rocks from five leg 60 sites. Init. Repts. of DSDP 60, (US Govern. Printing Office), 647-656.
47. Sharaskin A.Ya., Bogdanov N.A. and Zakariadze G.S. 1981. Geochemistry and timing of the marginal basin and arc magmatism in the Philippine Sea. Phil. Trans. R.Soc. London. A300, 287-297.
48. Sharaskin A. Ya., Dobretsov N.L. and Sobolev N.V. 1980. Marianites: The clinoenstatic bearing pillow-lavas associated with the ophiolite assemblage of the Mariana Trench. In: Proc. Int. Ophiolite Sym. Cyprus, 1979, 473-479.
49. Sharaskin, A.Ya., Karpenko S.F., Ljalikov A.V. et al., 1984. Correlated $^{143}Nd/^{144}Nd$ and $^{87}Sr/^{86}Sr$ data on boninites from Mariana and Tonga Arcs. "Ofioliti", Bull. del Gruppo di Lavoro sulle Ofioliti Mediterranee (in press).
50. Sharaskin A.Ya. and Zakariadze G.S. 1982. Peculiarities of the Philippine Sea floor magmatic evolution. In: Magmatism and metamorphism as indicators of the island arc geodynamics. Nauka, Moscow, 210-221 (in Russian).
51. Sharaskin A.Ya., Zakariadze G.S., Dmitriev Yu.I. et al. 1980. Petrology of the volcanic rocks. In: Geology of the Philippine Sea floor. A.V.Peive (Ed.), Nauka, Moscow, 106-148.
52. Shiki T., Aoki H., Suzuki H. et al. 1974. Geological and petrological results of the GDP Cruise in the Philippine Sea. Marine Sci. Monthly 6, N 8, 51-56.
53. Toksös M.N. and Hsui A.T. 1978. Numerical studies on back-arc convection and the formation of marginal basins. In: Toksöz M.N. (Ed.) Numerical Modeling in Geodynamics: Tectonophysics, 50, 177-196.
54. Vasiljev B.I. 1974. New data on the geology of the Amami and Borodino Plateau in the Philippine Sea.

In: Geology and geophysics of the N.-W. Pacific marginal Seas. Vladivostok, 122-127.

55. Watanabe T., Epp B., Uyeda S. et al. 1970. Heat flow in the Philippine Sea. In: Island arc and ocean. M. Hoshino and H.Acki (eds.). Tokai Univ. Press, 61-84.
56. Watanabe T., Langseth M.G. and Anderson R.N. 1977. Heat flow in back-arc basins of the Western Pacific. In: Talwani M., and Pittman W.C. III (Eds.). Island Arcs, Deep Sea Trenches and Back-arc Basins: Washington (Am. Geophys. Union), pp.137-161.
57. Watts A.B., Bodine J.H., and Bowin C.O. 1978. Free-air gravity field. In: Geophysical Atlas of East and Southeast Asian Seas, Hayes D.E. (ed.) Geol. Soc. Am. Map and Chart Series, MC-25.
58. Watts A.B., Weissel J.K. and Larson R.I.1977. Seafloor spreading in marginal basins of the western Pacific. Tectonophysics, 37, 167-181.
59. Wood D.A., Joron J.L., Marsh N.G. et al. 1980. Major and trace element variations in basalts from the North Philippine Sea drilled during Deep Sea Drilling Project Leg 58: a comparative study of back-arc basalts with lava series from Japan and mid-ocean ridges. In: Klein G., Kobayashi K. et al. Init. Repts. DSDP, 58, Washington (US Gov. Printing Office), 873-894.
60. Wood D.A., Mattey D.P., Joron J.L. et al. 1980. A geochemical study of 17 selected samples from the basement cores recovered at sites 447,448,449,450 and 451, DSDP Leg 59. In: Kroenke L, Scott R.B. et al. Init. Repts. DSDP, 59, Washington (US Gov. Printing Office), 743-752.
61. Yoshii T., Kono Y. and Ito K. 1976. Thickening of the oceanic lithosphere. IN: The Geophysics of the Pacific Ocean Basin and its Margin. Amer. Geophys. Union, Monogr. Ser. 19, 423-430.

FIGURE 1. Main structures of the Philippine Sea floor.
I - Basins, floored by the oceanic crust:
1) West Philippine, 2) Shikoky, 3) Parece Vela, 4) Bonin Trough, 5) Mariana Trough, 6) N.-W. Pacific, 7) West Caroline, 8) East Caroline.
II - Back-arc spreading centers:
1) Central Basin Fault (active from 50(?) to 38 Ma), 2)Kinan Seamounts, 3) IPOD Fracture Zone (both active from 30 to 18 Ma; spreading at the Mariana Trough started about 7 Ma).
III - Trenches as convergent plate boundaries:
1) Izu-Bonin, 2) Mariana, 3) Yap, 4) Palau, 5) Philippine, 6) Nansei-Soto (Ryukyu), 7) Nankai
IV - Active volcanic arcs:
1) Izu-Bonin (Iwo-Jima),2) Volcano, 3)Mariana, 4) S.-W.Japan, 5) Ryukyu, 6) Taiwan, 7) Philippine.
V - Remnant arcs:
1) Amami, 2)·Daito (both active in due Upper Mesozoic) 3) Palau-Kyushu (active from 40 to 29 Ma), 4) West Mariana (active from 20 to 9 Ma).
VI - Other ridges and rises:
1) Gogia Ridge, 2)Benham Rise, 3) Urdanet Plateau 4) Marcus Necker Seamounts, 5) Caroline Ridge, 6) Eauripic Ridge. (The inferred intervals of back-arc spreading and arc volcanic activity after (8,44,47)).

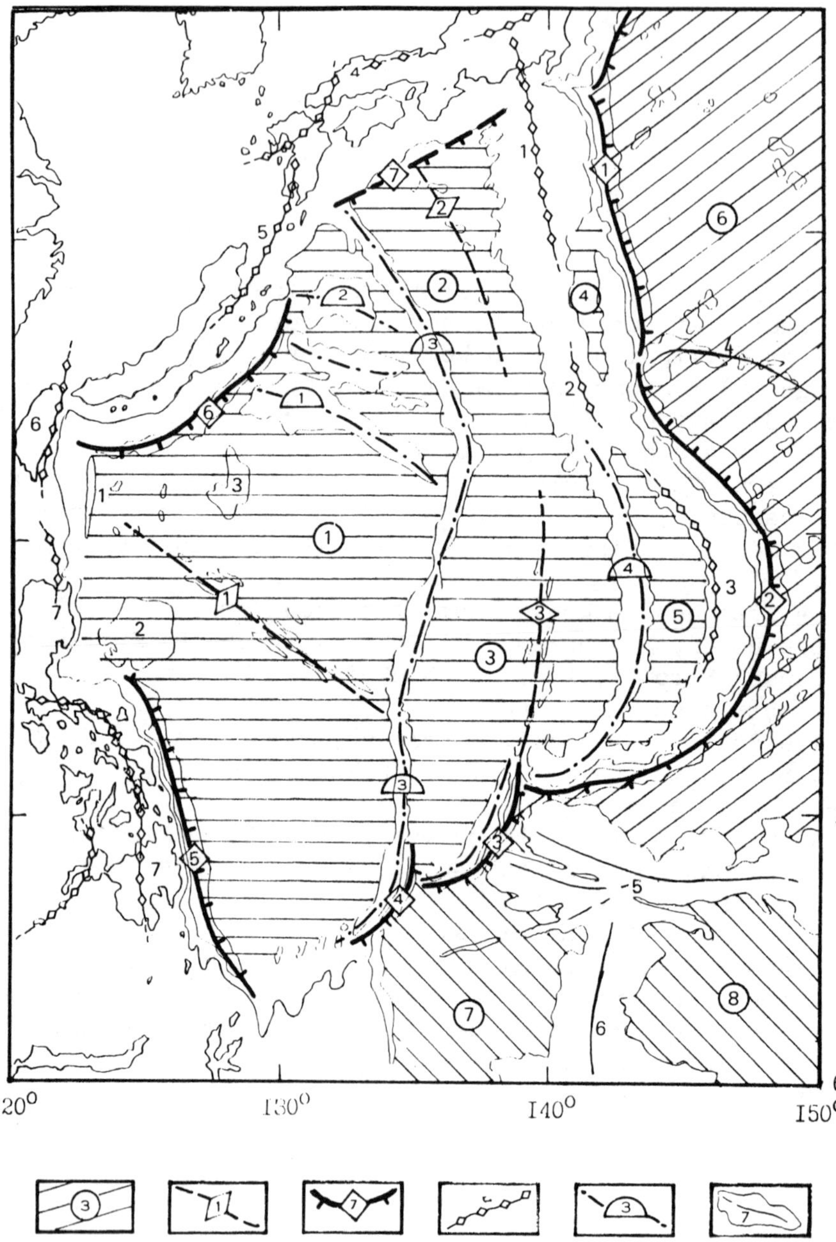
30°
20°
10°
0°
120°
130°
140°
150°

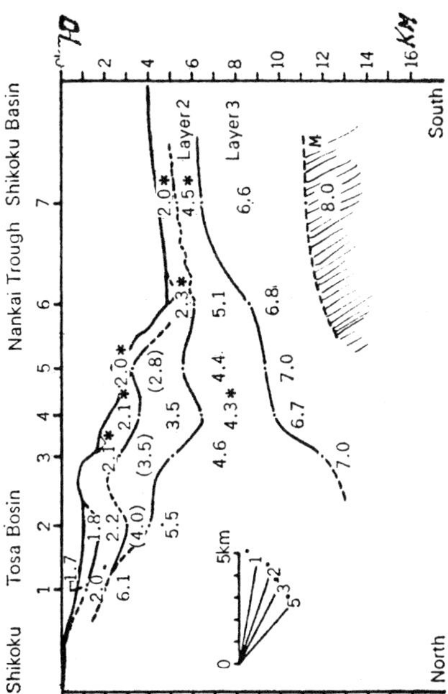
Shikoku
Tosa Bosin
Nankai Trough
Shikoku Basin
Layer 2
Layer 3
North
South
KM
5km

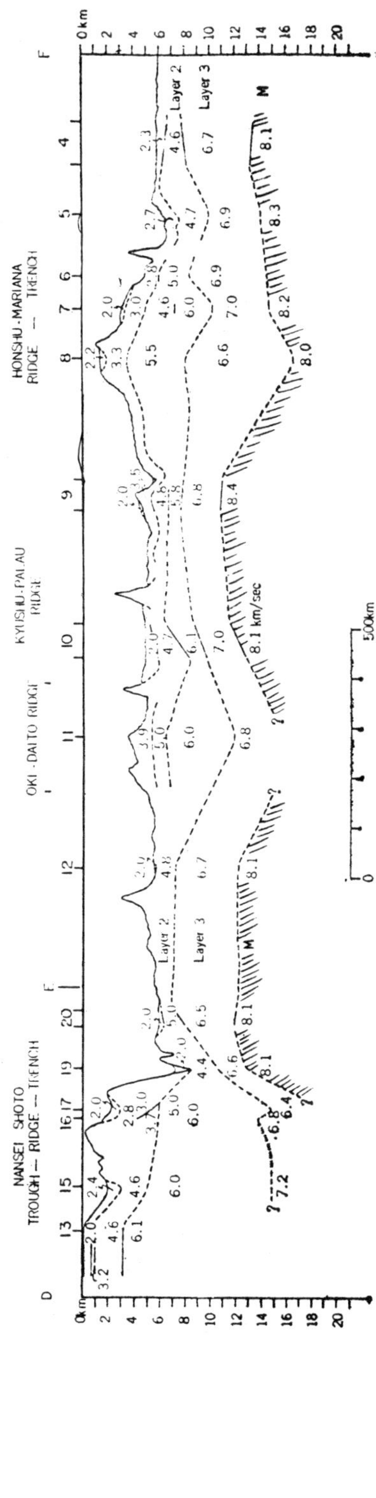
NANSEI SHOTO
TROUGH — RIDGE — TRENCH
OKI-DAITO RIDGE
KYUSHU-PALAU RIDGE
HONSHU-MARIANA
RIDGE — TRENCH
Layer 2
Layer 3
8.1 km/sec
500km
D
E
F

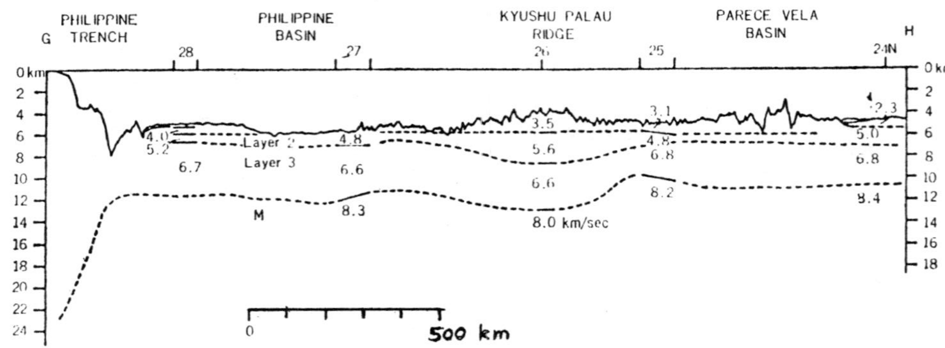

FIGURE 2. Crustal sections of the Philippine Sea floor (after (32,37); Seismic velocities in km/sec).

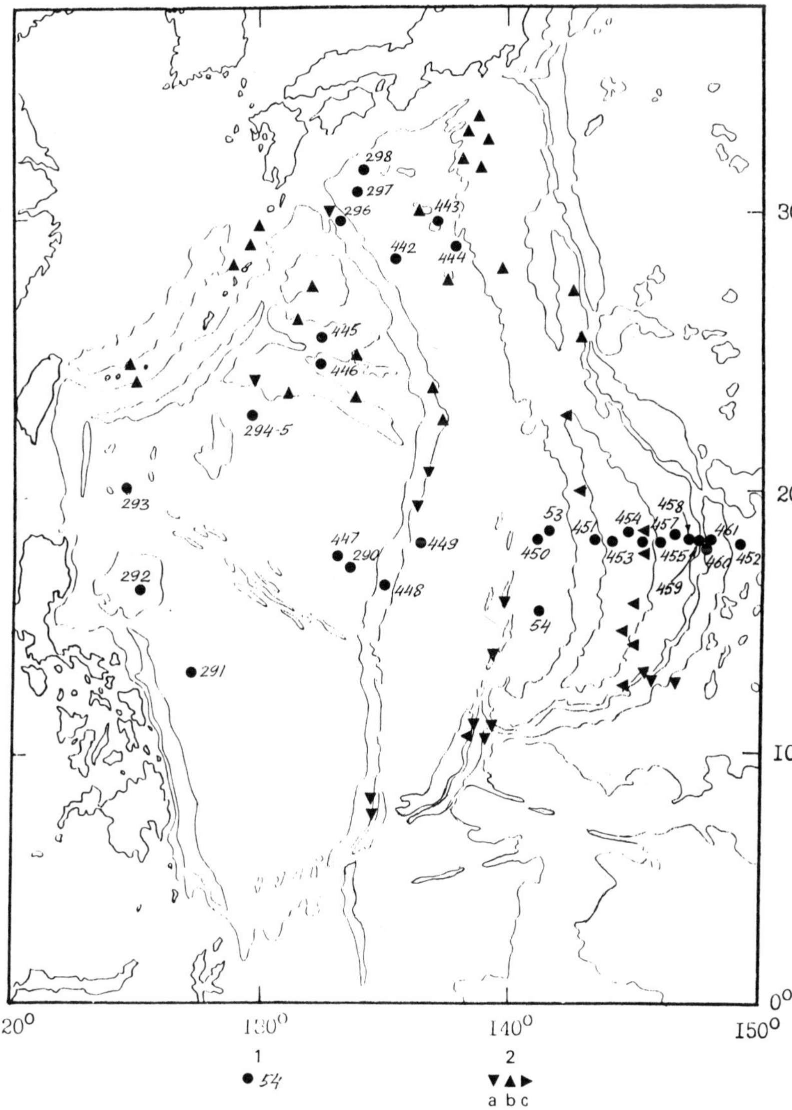

FIGURE 3. Locations of DSDP sites and dredging stations in the Philippine Sea (compiled after (11-13, 15,35,40,44, 54,52) and R.L.Fisher, personal communication). 1 - DSDP sites, 2 - dredging stations of: a - soviet, b - japanese and c - american marine expeditions.

FIGURE 4. Low and high heat-flow areas in the Philippine Sea (compiled after (3,36,43,55,56)).
The areas with predominating values less than 1.0 HFU=41.87 mWm^{-2} (deep shaded) are clinging to the trenches. The mean heat flow more than 1.5 HFU is characteristic to the large areas in the inner part of the sea (light shaded). Unshaded are areas of heat flow values from 1.0 to 1.5 HFU.

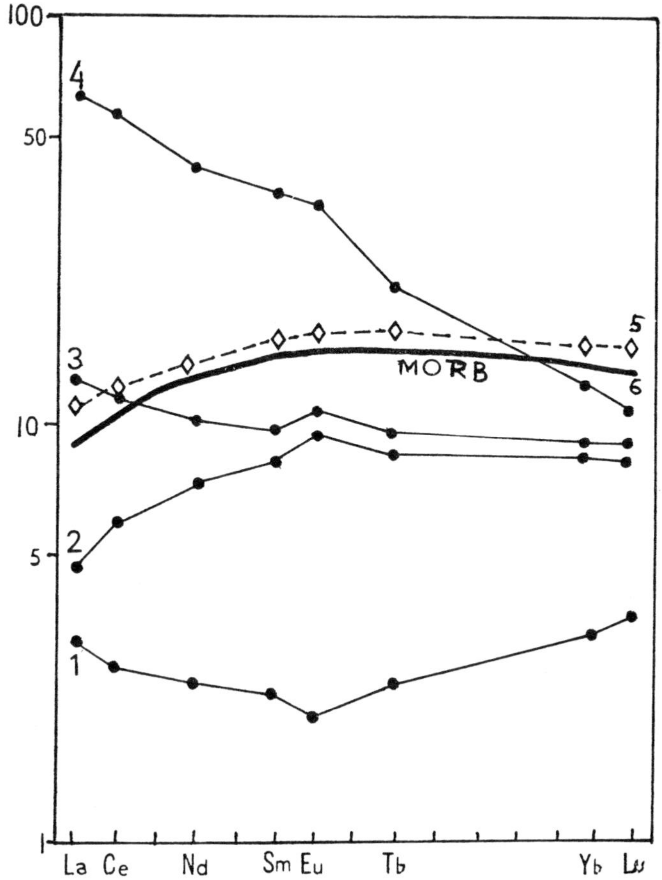

FIGURE 5. Chondrite normalized REE concentrations in various rock types of the Philippine Sea floor.
1 - boninite, 2 - arc tholeiite, 3- calc-alkali basalt, 4 - alkali basalt, 5 - inter-arc tholeiite, 6 - average mid-ocean ridge basalt (selected from (7,10,18,59,60)).

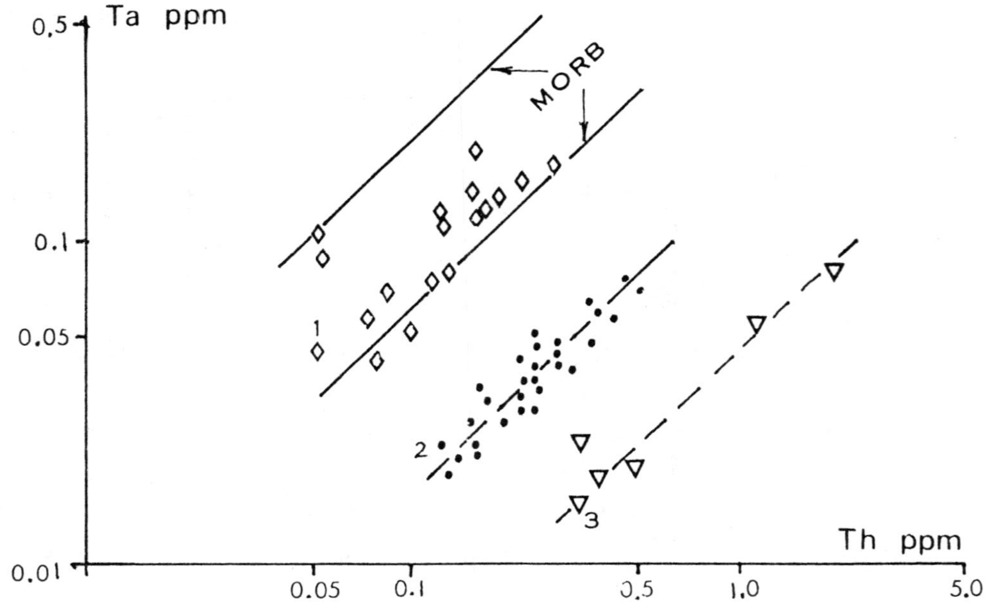

FIGURE 6. Ta versus Th correlation diagram for selected volcanic series of the Philippine Sea.
1 - inter-arc basin basalts, 2 - arc-tholeiite volcanic series, 3 - calc-alkaline volcanics, MORB - limits of Ta/Th ratios in different types of mid-oceanic ridge basalts (after (7, 59,60) and Zakariadze et al. unpublished data).

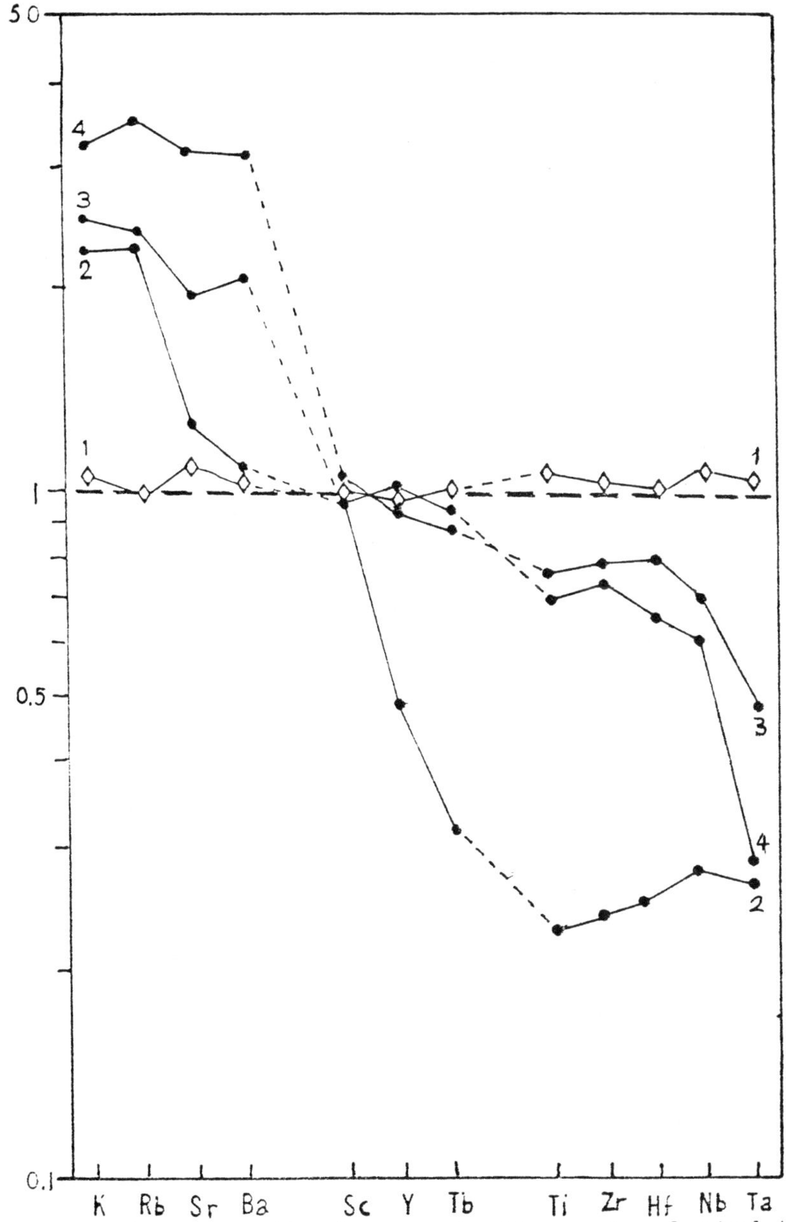

FIGURE 7. MORB normalized concentrations of selected trace elements in various rock types of the Philippine Sea.
1 - inter-arc tholeiite, 2 - boninite, 3 - arc tholeiite, 4 - calc-alkaline basalt (data sources as in Fig.5)

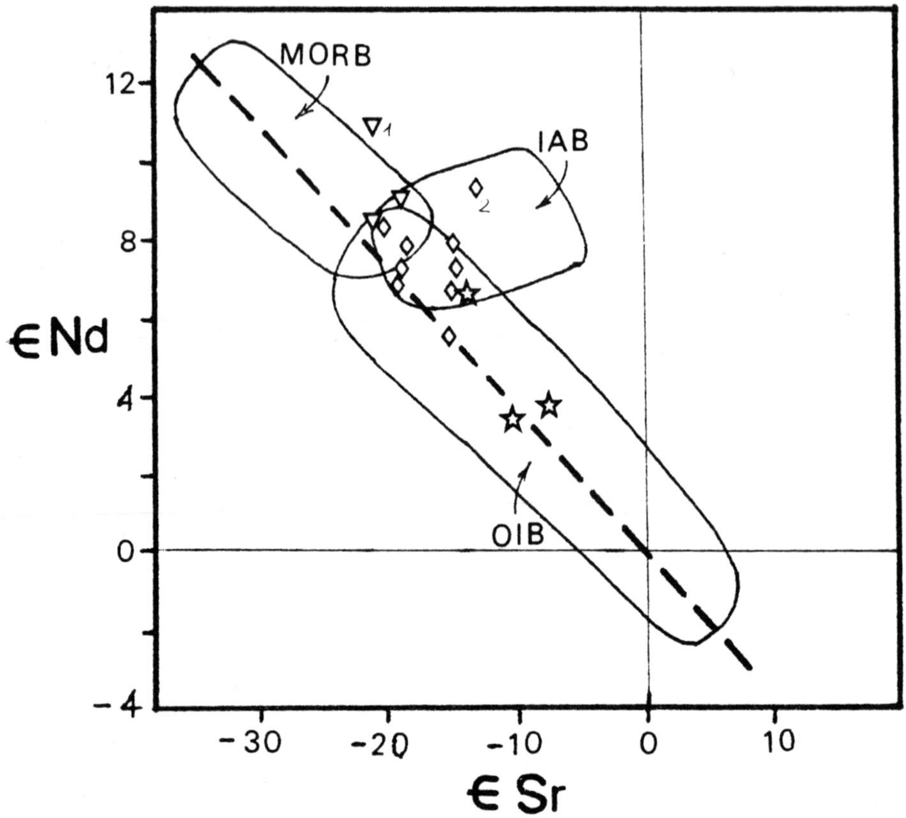

FIGURE 8. $^{143}Nd/^{144}Nd$ versus $^{87}Sr/^{86}Sr$ diagram for the volcanics of the Philippine Sea floor.
1 - inter-arc basin basalts, 2 - arc volcanics, 3 - boninites (after (9,49), MORB and OIB fields like in (9)).

THE RED SEA : A SMALL OCEAN BASIN FORMED BY CONTINENTAL EXTENSION AND SEA FLOOR SPREADING

COLEMAN, ROBERT G.
Geology Dept., Stanford University, California, U.S.A.

ABSTRACT

The formation of the Red Sea depression is considered to have developed in response to the separation of the Arabian plate away from the African plate. The first stages of separation probably began in the early Tertiary by incipient crustal extension that produced block faulting and a series of N-S trending morphotectonic depressions. By late Oligocene and early Miocene the main depression consisted of an extensive continental rift valley that was occupied by a large lake that received significant lacustrine sediments. Contemporaneous bi-modal volcanism within the rift valley produced interlayed pyroclastic and lake deposits along with significant accumulations of mafic and silicic flows. Crustal attenuation continued with the development of extensive dikes and subvolcanic intrusive complexes invading the stretched Pre-Cambrian crust. On the flanks of the Red Sea depression thick accumulation of plateau volcanics formed in Saudi Arabia, Yemen and Ethiopia. The left lateral displacement on the Dead Sea rift in early Miocene marks the first invasion of sea water into the Red Sea depression. Throughout the Miocene a thick evaporite blanket formed above the continental lacustrine and volcanic sequence as the depression continued to subside and extend. During the Miocene, geologic and geophysical observations are not conclusive, but there is very little evidence to support the idea that there was active seafloor spreading and formation of oceanic crust within the depression. At the beginning of the Pliocene (2-5 m.y.), sea floor spreading started in the middle portions of the Red Sea axial trough and continues up to the present time. Plate kinematic models of the Red Sea for the

Proceedings of the 27th International
Geological Congress, Volume 23, pp. 93–121
ORIGIN AND HISTORY OF MARGINAL AND INLAND SEAS

post-Pliocene are generally in agreement and fit nicely with reconstructed plate movements of the Middle East. Utilization of rigid plate kinematics or shoreline matching of the Red Sea where the basement for all of the depression is considered oceanic crust meets with serious geologic objections. Listric faults that rotate blocks of pre-Cambrian drust contemporaneous with bimodal igneous activity provide structural evidence that continental extension was active in the pre-Pliocene history of the depression. The unexplained overlaps in the northern and southern parts of the Red Sea are best explained as areas of attenuated continental crust developed during pre-Pliocene extension. Confining ocean crust formation to the narrow axial trough within the medial portions of the Red Sea overcomes nearly all of the previous rigid plate kinematic problems.

INTRODUCTION

The geologic setting of the Red Sea has fascinated geologists and geophysicists ever since Wegner's (1924) original description of the shoreline fit between Africa and Arabia (fig.1). The plate tectonic protagonists have treated the Red Sea in a number of general papers and have generally concluded that it represents a newly formed ocean developed as Arabia moved away from Africa (McKenzie et al., 1970; Girdler and Styles, 1974; Le Pichon and Franceteau, 1978). Nearly all of these workers agree with the general scenario of development but there remains an area of disagreement as to when the Red Sea actually opened and how much of the Red Sea is underlain by new oceanic crust. Geologic evidence seems to indicate that crustal attenuation began during the late Cretaceous and that the Red Sea originated as a depression or series of depressions receiving both marine and continental sediments (Coleman, 1974).

Geophysical evidence used in solving the kinematic relationships between the plate movements of Arabia -Nubia - Somalia have produced conflicting results. McKenzie et al. (1970) assumed a coast-line fit above 15°N in the Red Sea and calculated the pole of rotation to be 36.5°N, 18°E, and that all of the Red Sea depression was produced by sea floor spreading. Overlap of the Yemen on to the Afar area of East Africa in the pre-rifting shoreline fit was not satisfactorily explained and the actual separation was referred to as continental splitting. This splitting was not clearly described but presumably a fault was thought to extend through the crust at the beginning stages of spreading. Using magnetic data from the southern Red Sea, Girdler and Styles (1974) refined the kinematic model of McKenzie et al. (1970) and suggested that the Red Sea had a two stage history of sea floor spreading. By comparing synthetic magnetic profiles for these latitudes they were able to show an early phase of spreading during the late Eocene-early Oligocene (41-34 m.y.) followed by a second phase beginning 4 to 5 m.y. ago. Both McKenzie and Girdler used the seismic studies of Tromontini and Davies (1970) as the most important indirect evidence that the main trough of the Red Sea was underlain by oceanic crust. In contrast to these plate tectonic kinematic studies, a geological interpretation of the Red Sea's evolution by Lowell and Genik (1972) shows considerable faulting and continental attenuation with ocean crust occupying only the axial trough. Using exploratory well data as well as geophysical and geological information for the southern Red Sea they concluded that only the axial trough is underlain by oceanic crust and that the main trough consisted of Miocene-Pliocene sediments unconformably overlying Precambrian crust that had been thinned by listric faulting. They interpreted the seismic data of Tromontini and Davies (1970) on the main trough crust as indicating velocities comparable to a rifted Precambrian crust containing screens of new igneous crust. My earlier account of the Red Sea geology (Coleman, 1974) generally confirmed the model of Lowell and Genik. Volcanic and tectonic activity beginning in the Oligocene along the Red Sea and Afar depression marks the beginning of extensive continental attenuation (Mohr, 1978,1983; Coleman et al., 1977). Gass (1970) has interpreted this early volcanism as part of a large dome or swell centered over Yemen-Ethiopia-Somalia triple junction. However the nature and extent of the volcanism cannot be related directly to only sea floor spreading. Widespread eruption of alkali olivine basalts and silicic volcanics attest to deep seated mantle magma generation as well as partial melting of the crust. The large amount of new data on the Afar depression by the German, Italian and French teams has provided a link to the evolution of the Red Sea (Pilger and Rösler, 1976; CNR-CNRS, 1973). These studies reveal a history of crustal extension combined with widespread volcanic activity within the Afar rift system. As in the Red Sea, there remains a strong controversy as to the nature of the crust underlying the Afar depression. The seismic studies reveal intermediate velocities under the depression and these have been interpreted in various ways (Pilger and Rösler, 1976). The important concept of crustal attenuation proposed by Morton and Black (1976) for the Afar depression is supported by the extensive exposures of faulted and rotated pre-Miocene rocks that represent brittle extension above the plastic stretching of the Afar crust. The evidence of crustal extension across the Afar is supported, in part, by geophysical studies of the crust (Pilger and Rösler, 1976). The post-rift volcanics and sediments cover the central portion of the Afar and so only indirect methods can be utilized in estimating the actual amount of oceanic crust versus continental crust. Le Pichon and Franceteau (1978) attempted a new plate kinematic interpretation of the Red Sea - Gulf of Aden area after much of the work discussed above had been published. They pointed out that development of most continental margins involves subsidence and creep of the lower part of the continental crust. Thus a transitional zone may form by extension of the continental crust prior to the establishment of oceanic spreading. A reasonable kinematic analysis of the plate tectonic development of the Red Sea area by Le Pichon and Franceteau (1978) shows that using the McKenzie pole of rotation the angle of opening of the Red Sea may be 3.25° to 4.3° giving 200 to 150 km of new surface developed with at least 65 to 85 km of this formed since the Pliocene. They leave open the question of how the early surface was formed i.e. new crust or extension of Precambrian crust. The most recent analyses of Red Sea evolution by Cochran (1981,1983) using early plate kinematic methods combined with diffuse extension of continental crust provides a new model that seems consistent with both geophysical and geological data. Cochran accepts the McKenzie pole of rotation (36.5°N 18°E) and allows 4.34° of angular opening. Since the motion of the Arabian plate is controlled both by opening histories in the Gulf of Aden and Red Sea, Cochran has adopted a period of 10 m.y. forspreading in the Gulf of Aden as the parameter controlling the amount of opening in the Red Sea. From this assumption it is clear that movement on the Dead Sea fault, to the north, represents a plate boundary and that development of the Gulf of Suez is also a result of diffuse extension as well as in the northern part of the Red Sea. The important discovery that the sinistral strike slip movement on the Dead Sea rift is confined to the last 19-22 m.y. restricts the actual Red Sea opening to this post-Oligocene period (Bartov et al., 1980). Borrowing from recent studies and models on "Atlantic Type" passive margins it is Crochan's contention that a period of diffusive extension preceeds all sea-floor spreading along continental margins. Such a model can explain initial Red Sea basin subsidence and escarpment uplifts as part of the thermal perturbations in the asthenosphere as it rises and cools under the basin. In oppostion to Cochran's model a deep-refraction seismic profile across the eastern margin of the Red Sea by the USGS has been interpreted as confirming the presence of oceanic crust from shoreline to the axial trough of the Red Sea (Healy et al., 1982). This study found on the coast a transitional crust of 16 ± 4 km thins to about 8 km over the main trough of the Red Sea and has been interpreted as being typical of the oceanic layers 2 and 3. Furthermore, the diffuse magnetic anomalies recorded over the main trough are considered to represent magnetic anomalies produced by oceanic crust underlying the Miocene-Oligocene sediments (Healy et al., 1982; Girdler and Styles, 1974).

Even after this new data, the nature of the basement configuration under the Red Sea main trough still remains controversial. All exploratory wells have failed to penetrate into unequivocal old Precambrian basement or new oceanic crust within the main trough of the Red Sea, but as Cochran (1983) points out several wells near the shoreline have intersected Precambrian granite. Interpretation of the geophysical results do not provide unique answers and, as a result, a clear consensus on the early stages of the Red Sea's formation is elusive.

NATURE OF THE RED SEA DEPRESSION

The Red Sea occupies a narrow elongate depression that extends over 2000 km long but whose width varies from 180 km to 360 (Fig.2). Its shape is unusual, but in some ways is very similar to the Gulf of California both in shape and in origin. The main trough of the Red Sea is characterized by broad areas less than 1000 m deep and with a smooth topography. At thesouth end, flat lying Archipelagos of Farasan and Dahlak rise about 60 m above sea level and represent areas of salt diapir uplift. The Red Sea narrows to 28 km at the straits of Bab al Mandeb where it connects with the Gulf of Aden. Just north of the straits the axial part of the Red Sea is marked by a series of volcanic islands extending southward some 230 km from Jabal al Tair to Hanish Island.

The bathymetric chart of the Red Sea reveals a very distinct broad main trough extending northward from Hanish Island to the southern tip of the Sinai Peninsula (Laughton, 1970). A narrow axial trough (5-30 km wide) bisects the main trough throughout its entire length except for the southern segment south of Hanish Island. This axial trough has steep-sided walls with an extremely irregular bottom topography. Parallel faults apparently formed these steep sided walls and form the trough boundaries from Jebel Tair northward to lat. 25°N (Fig.2). Nortward of this point the axial trough is a broad U-shaped depression without prominent fault scarps. The main portion of the axial trough could be characterized as a graben with a central rough zone that has noticeable axial highs similar in some ways to the mid-Atlantic rift zone (Bäcker et al., 1975). Conical shaped edifices within the axial trough and on its flanks are considered to be single volcanic seamounts that extend southward from 24°N lat. to Jebel Tair where these volcanic cones rise above the surface and form a series of islands extending on south to Hanish Island. The discovery of hot brine pools overlying Fe-Mn metalliferous muds enriched in Zn, Cu, Pb, Ag and Cd within deeps (1500-2850 m) along the axial trough has promoted extensive oceanographic studies (Degens and Ross, 1969; Backer, 1975; Hilpert,1977; Ross et al., 1973). There are more than 19 occurrences of brine pools and/or metalliferous sediments between lats. 25°N and 19°30'N and the Atlantis II deep alone may contain as much as 150-200 million tons of mud that contain an average of 5% Zn and 1% Cu (based on dray and salt free samples) (Amann et al., 1973). Observations on black smokers within the east-Pacific rise and the Juan de Fuca ridge provide an actualistic tie to the formation of the Red Sea metalliferous muds (East Pacific Rise Study Group, 1981). The hot brine areas of the Red Sea are restricted between 19° and 24° and are related to areas of submarine volcanic activity produced by spreading in this part of the axial trough.

The Red Sea shelf has a bottom topography that is shallow (usually less than 10 meters) and is flat. South of 21°N lat. the Red Sea shelf merges imperceptibly with broad coastal plains on both sides whereas to the north these plains narrow and are cut by faults producing uplifted coral strands overlying marine and continental clastics. In a general way, it could be said that the southern Red Sea coastal plains are undergoing a rapid submergence whereas those to the north are emergent (Coleman,1974). This submergence has buried most of the structural manifestations of crustal attenuation and has led some authors to consider the Red Sea boundaries as monoclinal warps unrelated to faulting (Whiteman, 1968; Coleman, 1974).

Inland from the coastal plain on the Saudi Arabia side of the Red Sea an escarpment extends southward from 21°N lat. and parallels the coastline approximately 80 to 100 km from the Red Sea coastal plain (Fig.2). It rises to 3,700 in Yemen and averages 2000 m in Saudi Arabia. In Ethiopia and Sudan a similar scarp begins at 17°30'N lat.and reaches altitudes of over 3000 m in Ethipia where it turns southward to form the western flank of the Afar depression These escarpments apparently have developed during the rapid downwarping and thermal adjustment of the asthenosphere during extension similar to that described on the margins of other rifted sedimentary basins (Beaumont et al., 1982; Bott,1982).

The narrow coastal plains north of 21°N lat. are not bordered by matching escarpments such as are seen in the south. Instead, the bounding areas are cut by normal and lystric faults that parallel the Red Sea and develop a small scale basin and range topography partially covered by interdigitating fans and pediments along the coastal plain. The northern end of the Red Sea bifurcates into the Gulf of Suez and the Gulf of Elat, the former controlled by block faulting common to the northern Red Sea (Garfunkel and Bartov, 1977) and the later is controlled by a series of left lateral wrench faults extending northeastward beyond the Dead Sea (Ben-Avraham, 1983). In the northern area, the narrow coastal plains of the Red Sea abut against fault bounded blocks of basement less than 5-10 km from the shoreline and this topography illustrates that crustal extension must have continued well beyond the present Red Sea depression.

The Red Sea depression has a northern geomorphic expression quite distinct from the southern part and must be related to the degree of crustal attenuation versus spreading. Furthermore the matching of shorelines both in the south and in the north are imperfect and require unacceptable overlap of both the Tiranian Peninsula and Yemen. Cochran (1983) has shown that at least part of these overlaps can be accomodated if crustal attenuation is considered part of the Red Sea's opening.

SEDIMENTS WITHIN THE RED SEA DEPRESSION

The marine Mesozoic sediments within the vicinity of the Red Sea show no strong paleogeographic evidence that the depression in its present form existed during this time. (Coleman, 1974). Paleogeographically reconstructed Mesozoic marine shorelines cut directly across the present NW trend of the Red Sea and evidence of related tectonic activity during this time is generally lacking. Stratigraphic evidence points to the fact that the Red Sea developed mainly during Tertiary times. Some authors point to pre-Tertiary structures as controlling the development of the Red Sea but none of these seem important or possible to confirm.

The presence of early Tertiary sediments are reported in a number of areas around the Red Sea. During Eocene and Paleocene times, transgressive marine deposits developed in the northern Sinai and Egypt and a narrow gulf may have extended as far south as Sudan (Whiteman, 1968; Garfunkel and Bartov, 1977). On the Arabian side a similar narrow gulf paralleling the Red Sea trend extended from Jordan southward to south of Jeddah. These early Tertiary sediments on the Arabian side are mainly shallow water marine sediments and contain reworked laterite material that produces ironstones (Al-Shanti, 1966; Madden et al., 1979). No major structures control these sediments but the southward gulf-like shallow marine seas may have filled depressions that resulted from broad NW trending block faulting across Eastern Egypt, Sinai, Saudi Arabia, and Jordan. Eocene shorelines extend E-W across the southern part of the Red Sea and by the beginning of Oligocene the horn of Africa was completely emergent (Mohr, 1962).

Marine sediments of Oligocene age are unknown along the flanks of the Red Sea and exploratory deep oil well drilling has failed to encounter marine sediments of this age (Coleman, 1974). The southern Red Sea area near shore to Ethopia the Dogali series (Fig.3) of late Oligocene-early Miocene age consists of continental clastics interbedded with volcanic flows and pyroclastics (Hutchinson, and Engles, 1970). On the Saudi Arabian side a deep exploratory well (Fig.2) 25 km north of Jizan bottomed out in unfossiliferous clastic sediments consisting of shales, silts and sandstones and interbedded conglomerates (Gillman, 1968). These can be correlated with the Hammit Formation in Sudan which consists of continental clastics and interbedded basalt flows that overlie the Mukawwar formation of Paleocene-Upper Cretaceous age (Sestini, 1965). In the northern Red Sea and the Gulf of Elat continental deposits of lacustrine and fluvial origin (Rahman Formation) are deposited directly on Precambrian and are thought to represent closed morphotectonic basins developed during the late Oligocene or early Miocene (Garfunkel et al.,1974). A recent important discovery by Schmidt et al. (1982) of Oligocene-Early Miocene lacustrine sediments that contain vertebrates, fresh water fish in the Baid Formation are associated with bi-modal volcanics (Jizan Group) invaded by the Tihama-Asir volcanics of Oligocene-Miocene age. This new stratigraphic evidence demonstrates that during the Late Oligocene-Early Miocene the Red Sea depression began to recieve continental sediments within closed fluvial basins not unlike the present day east African Rift system or Basin Range of Nevada (Madden et al , 1979). Further work is required to confirm this Oligocene-Miocene continental rift stage of the Red Sea depression; however these stratigraphic correlations seem to preclude the idea that a Oligocene dome occupied the Red Sea depression at this time. Schmidt et al. (1982) exclude the possibility that these Oligocene-Miocene lacustrine sediments, continental clastics, and associated bi-modal volcanics (Jizan Group) extending much beyond the present shorelines into the main Red Sea depression. An exploratory oil well (B-1, Fig.2) in the southern Red Sea just north of the Dahlak islands encountered lacustrine argillites and basalts thought to be middle Oligocene by K/Ar measurements. Lowell and Genik (1972) considered these lithologies in B-1 to indicate a continental environment and equated these with the Dogali series exposed along the Ethiopian coastline. These late Oligocene-early Miocene continental clatics and lacustrine sediments imply widespread closed inter-montane basins developed as a result of early block faulting and continental extension. This new stratigraphic evidence points to the fact that the Red Sea depression received continental clastics and fluvial sediments in the early stages of continental attenuation rather than being part of a large Arabian-African dome (Gass, 1970).

Sometime during the early Miocene the Red Sea depression was invaded by marine waters from the Mediterranean and a thick series of evaporite deposits formed (Heybroek, 1965). More than 20 exploratory wells have been drilled in the Red Sea and so the extent and thickness of the salt is quite well known (Coleman, 1974). There are strong marked lateral changes in salt thickness between wells with interlayering of volcanogenic materials and continental clastics. The thickness varies from 1-4 km with only moderate thinning from the present shoreline towards the axial trough. B-1 well within few km of the axial trough shows more than 2 km of salt that rests on Oligocene basalt (Lowell and Genik, 1972). Detailed studies of the evaporites reveal that they developed in shallow water or sabkhah-type environments (Stoffers and Ross, 1977). Leg 23 (Ross et al., 1973) studies show that evaporitic and marine conditions fluctuated and that the source of the ocean water was from the Mediterranean where a nearly similar stratigraphic sequence has been recorded (Hsü and Bernoulli, 1978). Exposures around the Red Sea and from nearshore well data indicate the continental clastic sediments interlayered with the evaporites (Gillman, 1968;Sestini, 1965). The period of evaporite deposition extended throughout the Miocene at which time diffuse extension must have occurred throughout the Red Sea basin. At about 10 m.y., the Gulf of Aden began to open and the Dead Sea fault accomodated this extension to the north (Cochran, 1983). At the beginning of the Pliocene a major unconformity developed (Fig.3) in the Red Sea sedimentary sequence (Ross and Schlee, 1973). It is marked by the appearance of marine sediments that are characteristically nanno oozes varying from silty clays, chalk, and dolomites that contain pelagic fossils of Indian Ocean affinities (Stoffers and Ross, 1977). Seismic-reflection profiles show a widespread truncation of reflectors in the marine oozes above the evaporite section which has been interpreted as a major Pliocene-Pleistocene unconformity (Ross and Schlee, 1973). This unconformity is present throughout the main trough but does not extend into the axial trough. The lack of Miocene salt deposits in the axial trough has been established by drilling on leg 23 of the Deep Sea Drilling Project (Ross et al., 1973). In the axial trough a thin veneer of sediments is often interlayered with metalliferous deposits in the closed deeps associated with hot brines (Bäcker et al., 1975). The sediments in the axial trough are Pliocene or younger but in some instances salt flowage has invaded the axial trough and disrupted original stratigraphy (Girdler and

Whitmarsh, 1974).

Rates of deposition of Pliocene and Pleistocene sediments in the southern part of the Red Sea were distinctly higher than those observed at the latitude of Jeddah, indicating continued uplift and erosion along the southern margins of the Red Sea (Ross et al., 1973). On the Saudi Arabian coast the Bathan Formation of Pliocene age consists of continental clastics and overlies the earlier formed Tihama-Asir volcanic complex and is thought to mark the uplift of the escarpment (Schmidt et al., 1983) (Fig.3). Within the Mansiyah well Gillman (1968) described a sequence of continental clastics more than 2000 meters thick overlying the slat and they have sandstone conglomerate at the base and extend upward into shale, shabka deposits, and volcanic flows. On the Sudan coast Sestini (1965) reports a Pliocene clastic group of very coarse gravels resting on Miocene strata and relates this to intense faulting and uplift during the middle or late Pliocene (Hutchinson and Engels, 1970). Within the Afar depression the Danakil - Red Series represent a clasitic sequence related to uplift and has been dated as Mio-Pliocene in age (CNR-CNRS, 1973). All of these stratigraphic indicators provide clear evidence that profound changes accompanied the actual beginning of rifting within the axial trough which started in early Pliocene (Fig.3).

The Holocene coral reefs along the Red Sea often uplifted and deformed in the north, whereas to the south continued sinking of the coral reefs indicates that they are being rapidly covered by sediments as the Red Sea depression continues to subside.

TECTONICS AND VOLCANISM OF THE RED SEA

Considerable new work has been published on this subject during the last ten years although not all of the authors are in agreement with the general sequence of events. As was shown with sedimentary sequence, most of the Red Sea's history is confined to the Tertiary and this also seems to be true for the associated volcanic rocks. There are four main areas of volcanic activity involved in the evolution of the Red Sea: 1. Ethiopia (Afar depression); 2. Yemen plateau; 3. Western Saudi Arabia; 4. Axial trough of the Red Sea. All of these areas contain volcanic sequences that relate directly to the development of the Red Sea.

The Ethiopian volcanics can be divided into two main groupings (Mohr, 1978; 1983): 1. The plateau tholeiitic basalts occurring between 36-15 m.y. and form thick piles (up to 3000 m) around the edges of the Afar. This eruptive period was followed by series of rhyolitic volcanics (15-11 m.y.) emplaced during diffuse extension of the Red Sea. 2. The Afar depression consists of a thick stratoid series of basalts often intercalated with rhyolites. The axial range of the Afar consists of oceanic basalts erupted from 1.2 m.y. up to the present, associated with marginal rocks of both silicic and volcanic nature. An early peralkaline granite series of Miocene age (22-25 m.y.) marks the beginning of igneous activity in the Afar and coincides with the granophyres along the Red Sea and the peralkaline intrusives of the Yemen.

It is possible to relate most of the Afar volcanic events with faulting that preceded eruption (Mohr, 1978) which resulted from crustal extension along a northwesternly trending axis. This process of volcanism and attenuation gave rise to a floor in the Afar depression that progressively sank and flanks that continued to rise. Different types of magmatism evolved simultaneously within the Afar depression as a result of distinct types of tectonism. Actual crustal separation taking place along the axial range gave rise to MORB basalts whereas along the flanks crustal attenuation prevailed and intermediate basalts formed, in other parts, partial melting of the continental crust by the ascending basalts produced silicic volcanics. In general then, the Afar depression and surrounding volcanics are synchronous with the development of the Red Sea depression and represent an on-land example of how the Red Sea evolved (Barberi et al., 1972; Pilger and Rösler, 1976).

In Yemen, details of its Tertiary history are not as well known, however the plateau volcanics are similar to those found in Ethiopia but at much less volume (Grolier and

Overstreet, 1978). These volcanics have a horst and graben structure showing tilting of individual blocks which are cut by many dikes related to extension. Here thicknesses of up to 1000 m of interlayered basalts and rhyolitic (1:1) ignimbrites form the imposing volcanic plateau that rises 3000 meters above the Red Sea coastal plain (Civetta et al., 1978). This volcanic pile is cut by a peralkaline granite (22 m.y.) similar to those found in the Afar (Grolier and Overstreet, 1978).

The Tertiary volcanic rocks of Saudi Arabia are more closely related to the tectonic evolution of the Red Sea. They represent one of the world's largest areas of alkali olivine basalt covering some 90.000 km^2. These basalts occupy 13 separate volcanic centers confined to the eastern coast of the Red Sea to the western Arabian highlands from Syria southward where they merge with the Yemen Plateau (Fig.2). The oldest basaltic lavas range from 29 to 18 m.y. and relate to the earliest stage of volcanic activity that has NW trending vent systems. These first (pre-uplift) basalts erupted on the Arabian plate are predominately picrites and ankaramites whereas dikes intruded near the axis of the proto-Red Sea rift zone are tholeiitic. These early basalts evolved from picrite-ankaramites to alkali olivine basalt. Minor volumes of fractionated, undersaturated felsic rocks are associated with the first stage rocks. In northern Saudi Arabia the presence of Eocene to Paleocene marine sediments as well as continental fanglomerate deposits beneath the basalt flows of these early eruptions suggests that block faulting and continental extension preceded this early volcanism. These extensional structures may have allowed gluf-like prongs of the Mediterranean.Tethys to reach as far south as Jeddah during late Cretaceous and early Tertiary. In the southern part of the Red Sea, the incipient lithospheric thinning during the Oligocene may have formed a series of northwest-trending lacustrine block faulted basins that accumulated fresh-water sediments and bimodal volcanic rocks in their central parts and continental clastics along their margins. The presence of vertebrate fossils in these siliceous lacustrine sediments indicates that the basins were similar to present day east Africa (Schmidt et al., 1982).

The eruptive Early Miocene late Oligocene sequences within the southern part of the Red Sea along its flanks where lystric faults developed as the crust attenuated were also bi-modal. These magmas were emplaced as dikes, sheets, layered bodies and flows that mark discontinuous zones of new igneous crust within the extending continental crust (Coleman et al., 1977). These dikes extend from Yemen to the Sinai Peninsula along a NW trend (Bartov et al., 1980) The abundance of rhyolite dikes, sheets, and granitic plugs and the absence of rocks of intermediate composition indicate that these silicic magmas are melts derived from subcrustal continental rocks partially melted by ascending tholeiitic magmas that later formed layered gabbros and dike swarms.

Following the initial extension of the continental crust in Early Miocene volcanism waned around the Red Sea just before middle Miocene times 16-14 m.y. in a fashion similar to that seen in the Afar depression. At this time, beginning left-lateral movement on the Dead Sea rift zone changed the stress pattern on the Arabian plate as it moved away from Africa (Bartov et al., 1980). A second stage of alkali olivine basalt extrusion commenced along north-trending fractures during the Pliocene, and recent volcanic activity in Saudi Arabia is also controlled by N-S fractures. The post-Pliocene extrusions form extensive areas of basalt from Jordar sothward to Yemen however the total volume is not significant when compared with the Oligocene to Early Miocene plateau lavas of Yemen and Ethiopia. In the post-Pliocene eruptions, a general eruptive sequence has been developed and applies to nearly all of these separate volcanic fields of Saudi Arabia. The basal flows are usually picrite-ankaramites followed by the alkali olivine basalts that locally contain mantle peridotite inclusions (Ghent et al., 1980). Diktytaxitic basalts containing cotectic mineral assembalges follow the more primitive rocks and they, in turn, are followed by hawaiites and mugearites with plagioclase megacrysts that generally represent the final stages of eruptive history. Minor but important bodies of silicic rocks are associated with this final sequence of eruption. The trachytes, phonolites, and comendites are believed to be differentiation products of hawaiites in high level magma chambers, or may possibly be partly mobilized and melted Precambrian crust (Baker et al., 1973).

The axial trough of the Red Sea is characterized by fissure eruptions that extrude on the inner floor within the deepest part of the graben. These flows form low hills and a rough topography. The only information on these rocks has been derived from dredging and from DSDP Site 226 (Coleman et al., 1974; Chase, 1969). This active volcanism seems to be confined between 22°N Lat. and 17°N Lat. In the northern part of the Red Sea axial trough, extrusive volcanic areas have not been established except where magnetic highs could indicate local intrusive activity. Chemical and petrologic studies of the axial trough basalts show them to be subalkaline tholeiites similar to mid ocean ridge basalts (MORB). They are hypersthene normative with initial Sr^{87}/Sr^{86} ratios and Pb^{206}/Pb^{207} values that are essential identical to MORB (Coleman et al., 1974). Where the axial trough narrows south of 16°N a series of volcanic islands occupy the axial trough beginning with Jabal Tair (Gass et al., 1973). It is a small shield volcano of known historic eruption producing subalkaline basalt tholeiitic basalt that rises 1200 m above the trough and has inumdated and covered the axial portion of the trough. Further south the Zubair Island group occupies a similar situation but consists of a series of lavas that are intermediate between tholeiitic and alkali basalt (Gass et al., 1973). Still further south the Zukur-Hanish volcano island group trends NE across the axial part of the trough and consist mainly of alkali olivine basalts accompanied by differentiates that are trachy-basalts, trachy-andesites, and trachytes (Gass et al., 1973). This dramatic change in volcanic rock type could mark a zone where diffuse continental extension spans the southern Red Sea, and no oceanic crust forms within an axial trough (Fig.2). The axial trough north of Jabal Tair contains many seamounts that consist of volcanic rocks produced from submarine eruptions. Little is known about the petrology of these submarine volcanic peaks but glasses dredged from some peaks indicate both tholeiitic and alkali basalt affinities (Bäcker et al., 1975). It is generally agreed that the axial trough extrusive tholeiites are related to present day spreading centers developed as a result of drift of Arabia away from Africa. The apparent confinement at these axial tholeiites between 17° to 22°NLat. and the progrssive change southward from active spreading to volcanic edifices astraddle the axial trough indicates a major change in subcrustal processes. It is interesting to note that on the Hall et al. (1977) map of the northern Red Sea the magnetic anomalies are confined mostly to ellipsoidal areas within the axial trough. These anomalies are not like linear anomalies normally associated with oceanic crust and it would appear that they most likely represent intrusive complexes similar to the Tihama Asir (Cochran, 1983). Extension in the northern part of the Red Sea probably has not advanced to a point where continuous spreading centers have developed.

CONCLUSIONS

The present day knowledge of the Red Sea has greatly improved over the past ten years. A plate tectonic kinematic model for the past about 5 m.y. has a general consensus whereby the medial part of the axial trough shows 50-60 km of new oceanic crust whose geophysical signatures are consistent and fit nicely with present day reconstruction of plate movement (McKenzie et al., 1970; Le Pichon and Franceteau, 1978). The history of the Afar region and the Gulf of Aden provides both geologic and geophysical evidence that they also participated in Red Sea plate movement and that their igneous events can be correlated (Mohr, 1978). The main difficulty arises in reconstructing the Pre-Miocene history of the area (Cochran, 1983). When rigid plate kinematics (or shoreline matching) are utilized to reconstruct plate movements the configurations clash with the known geologic relationships. The realization that both the lithosphere and asthenosphere react in a non rigid way during plate separation has provided new insights regarding the evolution of the Red Sea (Royden and Keen, 1980; Beaumont et al., 1982; Le Pichon and Sibuet, 1981). The presence of listric faults associated with rotated fault blocks in narrow and restricted zones along the Red Sea coastline provides evidence that these same structures must extend across much of the main trough (Coleman, 1983) (Fig.4). The disturbed block faulted nature of the upper surface of the Miocene salt deposits throughout the Red Sea indicates faulting continued after the deposition and that nearly all of these structures have been burried by post-Miocene sediments the Red Sea (Ross and Schlee, 1973). Comparison of similar buried margins along the Atlantic provides a striking analogy of development (Royden et al., 1980). The now exposed tectono-magmatic province of E Greenland that developed during the early opening of the Atlantic bears striking similarity to the Tihama-Asir complex on the Red Sea Coast and to much of the volcanic sequence in the Afar. The early production of bimodal volcanics followed by voluminous outpouring of plateau basalts marks the early rise of the asthenosphere during diffuse extension. Rapid depression of the margin and rise of the flanks marks isostatic adjustment to cooling and heating of the lithosphere (Beaumont et al., 1982). Slow rates of diffuse extension produce closed basin and salt deposits, whereas more rapid extension develops narrow zones of extension characterized by E Greenland (Brooks and Nielsen, 1982). Extremely slow rates of extension produce the East African Rift system or Gulf of Suez. Continued studies of the Red Sea should provide answers regarding the nature of continental rupture. Here we need to combine our knowledge of sedimentation, volcanology, tectonics, and the geophysical restraints of plate tectonic kinematics.

REFERENCES

Al-Shanti A.M.S., 1966; Oolitic iron ore deposits in Wadi Fatima between Jeddah and Mecca, Saudi Arabia, Saudi Arabian Directorate General of Mineral Resources Bull., 2, 51 p.

Amann H., Bäcker H., Blisserbach E., 1973, Metalliferous muds of the marine environment- offshore Tech. Cont. OTC 1759, p. 345-358, Dallas.

Bäcker H., 1975, Exploration of the Red Sea and Gulf of Aden during the M.S. Valdivia Cruises "Erzschlämme A" and "Erzschlämme B", Geol. Jahrb. v. D13, p. 3-78.
Bäcker H., Lange K., Richter H., 1975, Morphology of the Red Sea central graben between Subair Islands and Abul Kizaan, Geol.Jahrb., v.D13, p.79-123.
Baker P.E., Brosset R., Gass I.G., Neary C.R., 1973, Jebel al Abyad: A recent alkalic volcanic complex in western Saudi Arabia: Lithos, v.6, no. 3, p. 291-314.
Barberi F., Varet J., 1975, Recent volcanic units of Afar and their structural significance (A summary), in: A.Pilger and A.Rösler (eds.)Afar depression of Ethiopia, Schweizbart, Stuttgart, p.174-178.
Barberi F. and others, 1972, Geological Map of the Danakil Depression, 1:500.000 (Northern AFAR-ETHIOPIA) CNRS France, CNRS Italy.
Bartov Y., Steinitz G., Eyal M., Eyal Y., 1980, Sinistral movement along the Gulf of Aqaba - its age and the relation to the opening of the Red Sea: Nature, v.285, p.220-222.
Beaumont C., Keen C.E., Boutillier R., 1982, A comparison of foreland and rift margin sedimentary basins, Phil. Trans. R.Soc.Lond. A.v. 305, p. 295-317.
Ben-Avraham Zvi., 1983, Sructural framework of the gulf of Elat (Aqaba) northern Red Sea, Journ. Geoph. Res. (in press).
Bott M.H.P., 1982, Origin of the lithosperic tension causing basin formation, Phil.Trans. R.Soc. Lond.A.v.305, p.319-324.
Brooks C.K., Nielsen T.F.D., 1982, The E.Greenland continental margin: a transition between oceanic and continental magmatism. J.Geol.Soc. London, v.139, p.265-275.
Chase R.L., 1969, Basalt from the axial trough of the Red Sea, in: Degens, E.T. and Ross D.A., eds., Hot Brines and recent heavy metal deposits in the Red Sea: New York, Springer-Verlag, p.122-128.
Civetta L., La Volpe L., Liver L., 1978, K-Ar ages of the Yemen Plateau: Jour. of volcanology and geothermal Research, v.4, p. 307-314.
Cochran J.R., 1983, A model for development of Red Sea, Amer.Assoc. Pet. Geol. Bull. v.67, p.41-69.

Cochran J.R., 1981, The Gulf of Aden: Structure and evolution of a young ocean basin and continental margin, Jour. Geophys. Res. v. 86, p.263-287.
CNR-CNRS AFAR TEAM, 1973, Geology of northern Afar (Ethiopia), Rev. Geogr. Phys., Geol., Dynam. (2), v. 15, p. 443 - 490.
Coleman R.G., 1984, The Tihama Asir Igneous Complex, a passive margin ophilite, 27th Int.Congress Moscow, Sec. c., 091.2 submitted (1983).
Coleman R.G., 1974, Geological background of the Red Sea: Initial Repts.Deep Sea Drilling Project, v.23, p. 813-820.
Coleman R.G., Fleck R.J., Hedge C.E., Ghent E.D., 1977, The volcanic rocks of southwest Saudi Arabia and the opening of the Red Sea, in: Hilpert L.S. ed., Red Sea Research, 1970-1975, Saudi Arabian Directorate General of Mineral Resources, Bull., 22, p.D1-D30.
Coleman R.G., Tatsumoto M., Coles D.G., Hedge C.E., Mays R.E., 1974, Red Sea Basalts, EOS, Amer.Geophys. Union Trans. v. 54, no.11, p.1001-1002.
Drake C.L., Girdler R.W., 1964, A Geophysical study of the Red Sea: Geophys. Journ. Roy, Astro. Soc. v.8, No. 5, P. 473-495.
Degens E.T., Ross D.A. (eds.), 1969, Hot brines and recent heavy metal deposits in the Red Sea, Springer (New York), 600 p.
East Pacific Rise Study Group, 1981, Crustal processes of the mid-ocean ridge, Science, v.213, p. 31-40.
Garfunkel Z., Bartov Y., 1977, The tectonics of the Suez rift, Geol. Surv. Israel, Bull., 71, 44p.
Garfunkel Z., Bartov J., Eyal Y., Steinitz G., 1974, Raham conglomerate - new evidence for Neogene tectonism in the southern part of the Dead Sea Rift, Geol. Mag., v.III p. 55-64.
Gass I.G., 1970, The evolution of volcanism in the junction area of the Red Sea, Gulf of Aden and Ethipian rifts Phi:.Trans.Roy.Soc.Lond.A.v.267, p. 369-381.
Gass I.G., Mallick D.I.J., Cox K.G., 1973, Volcanic islands of the Red Sea: Jour. Geo.Soc. Lond. v. 129, p.275-310.
Ghent E.R., Coleman R.G., Hadley D.G., 1980, Ultramafic inclusions and host alkali olivine basalts of the south-

ern coastal plain of the Red Sea, Saudi Arabia, Amer. Jour. Sci. v.280-A, p.499-527.
Gillman M., 1968, Primary results of a geological and geophysical reconnaissance of the Jizan coastal plain in Saudi Arabia: Proc. 2 nd AIME Regional Technical Symposium Dhahran, p. 189-208.
Girdler R.W., 1970, A Review of Red Sea heat flow: Phil. Trans.Roy.Soc.Lond. A.v. 267, p.191-203.
Girdler R.W., Styles P., 1974, Two-stage Red Sea floor spreading, Nature, v.247, p.7-11.
Girdler R.W., Whitmarsh R.B., 1974, Miocene evaporites in Red Sea cores, their relevance to the problem of the width and age of oceanic crust beneath the Red Sea, Int. Repts. Deep Sea Drilling Project, v.23, p.913-922.
Grolier M.J., Overstreet W.C., 1978, Geologic map of Yemen Arab Republic (San'a), U.S.Geological Survey Misc. Inv. Series, Map 1-1143-B, Scale 1:500.000.
Hall S.A., Andreasen G.E., Girdler R.W., 1977, Total-intensity magnetic anomaly map of the Red Sea and adjacent coastal areas, a description and preliminary interpretation, in: L.S.Hilpert, ed., Red Sea Research 1970-1975: Saudi Arabia Div. Gen. Mineral Resources Bull., 22, p. F1-F15.
Healy J.H., Mooney W.D., Blank H.R., Gettings M.E., Kohler W.M., Lamson R.J., Leone L.E., 1982, Saudi Arabian seismic deep refraction profile: Final Project Report. Saudi Arabian Deputy Ministry for Mineral Resources Open File Report USGS-OF-02-37, 141 p.
Heybroek F., 1965, The Red Sea Evaporite Basin, in: Salt Basins Around Africa: Inst. Petroleum, p.17-40, London.
Hilpert L.S., (ed.), 1977, Red Sea Research 1970-1975, Saudi Arabia Div. Gen.Mineral. Resources, Bull., 22.
Hsü K., Bernoulli D., 1978. Genesis of the Tethys and the Mediterranean, in K.J.Hsü, L.Montadert et al., Int. Rept. Deep Sea Drilling Project v. 42, Pt1 Wash.D.C., U.S., Govt. Printing Office, p. 943-949.
Hutchinson R.W., Engels G.G., 1970, Tectonic significance of regional geology and evaporite lithofacies in northeastern Ethiopia: Phil.Trans. Roy.Soc.Lond. A. v. 267, p. 313-329.
Laughton A.S., 1970, A new bathymetric chart of the Red

Sea : Royal Soc.(London) Philos. Trans. v. 267, no.1181, p. 227-266.

Le Pichon X., Francheteau J., 1978, A plate tectonic analysis of the Red Sea, Gulf of Aden area: Tectonophysics, v. 46, p. 369-406.

Le Pichon X., Sibuet J.C., 1981, Passive margins: a model of formation, Jour. Geophys. Res., v. 86, p.3708-3720.

Lowell J.D., Genik G.J., 1972, Sea Floor spreading and structural evolution of southern Red Sea, Amer.Assoc. Pet. Geol.Bull., 56, p.247-259.

McKenzie D.P., Davies D., Molnar P., 1970, Plate tectonics of the Red Sea and East Africa: Nature, v.226, p.243-248.

Madden C.T., Nagri I.M., Whitmore F.C.Jr., Schmidt D.L., Langston W.Jr., Wood, R.G., 1979, Paleocene vertebrates from coastal deposits in the Harrat Hadan area, at Taif region, Kingdom of Saudi Arabia, U.S.Geological Survey, Saudi Arabian Mission, Proj. Report 269, 29 p.

Madden C.T., Schmidt D.L., Whitmore F.C.Jr., 1983, Masritherium (Artiodactyla, anthracotheriidae) from Wadi Sabya, southwestern Saudi Arabia: An earliest Miocene age for continental rift-valley volcanic deposits of Red Sea margin, Saudi Arabian Deputy Ministry for Mineral Resources, Open File Report, USGS-OF-03-61, 24 p.

Mohr P., 1983, Ethiopian flood basalt province, Nature, v. 303, p.577-584.

Mohr, P., 1978, Afar, Ann.Rev.Earth.Planet.Sci., v.6, p. 145-172.

Mohr P., 1962, The geology of Ethiopia: Addis Ababa Univ. Coll.Press.286 p.

Morton W.H. and Black R., 1975, Crustal attenuation in Afar, in: A.Pilger and A.Rösler (eds.), Afar depression of Ethiopia, Schweizbart, Stuttgart, p.55-65.

Pilger A., Rösler A., 1976, Afar depression of Ethiopia, Vol.I; Afar between continental and oceanic rifting, Vol. II: Stuttgart, E., Schweitzbart'sche Verlagsbuchhandlung.

Ross D.A.,Schlee J., 1973, Shallow structure and geologic development of the southern Red Sea, Geol.Soc.Amer. Bull. v. 84, p.3827-3840.

Ross D.A., Whitmarsh R.B., Ali S.A., Boudreaux J.E., Coleman R.G., Fleisher R.L., Girdler R., Manheim F., Matter A., Nigrini C., Stoffers P., Supko P.R., 1973, Red Sea drillings: Science, v.179, p.377-380.

Royden L., Sclater J.G., Von Herzeb R.P., 1980. Continental margin subsidence and heat flow: Important parameters in formation of petroleum hydrocarbons, Amer. Assoc. Pet. Geol. Bull., v.64, p.173-177.

Royden L., Keen C.E., 1980, Rifting processes and thermal evolution of the continental margin of eastern Canada determined from subsidence curves, Earth and Planetary Science Letters, v.51, p.343-361.

Schmidt D.L., Hayden D.G., Brown G.E., 1982, Middle Tertiary continental rift and evolution of the Red Sea in southwestern Saudi Arabia, Saudi Arabian Deputy Ministry for Mineral Resources, Open File Report USGS-OF-03-6, 56 p.

Sestini J., 1965, Cenozoic stratigraphy and depositional history, Red Sea Coast, Sudan : Bull., Amer.Assoc. Pet. Geol. v.49, p.1453-1472.

Stoffers P., Ross D.A., 1977, Sedimentary history of the Red Sea, in: Red Sea Research 1970-1975, Saudi Arabia Div.Gen.Mineral Resources, Bull., 22, p.D1-30.

Tapponier P., Francheteau J., 1970, Necking of the lithosphere and the mechanics of slowly accreting plate boundaries, JOur.Geophys.Res. v.83, p.3955-3970.

Tramontini C., Davies D., 1970, The deep structure of the Red Sea: Phil. Trans. Roy. Soc.London A. v. 267, p.181-189.

Whiteman A.J., 1968, Formation of the Red Sea depression: Geol. Mag, v.105, p.231-246.

Wegener E., 1924, The origin of continents and oceans, London Methuen.

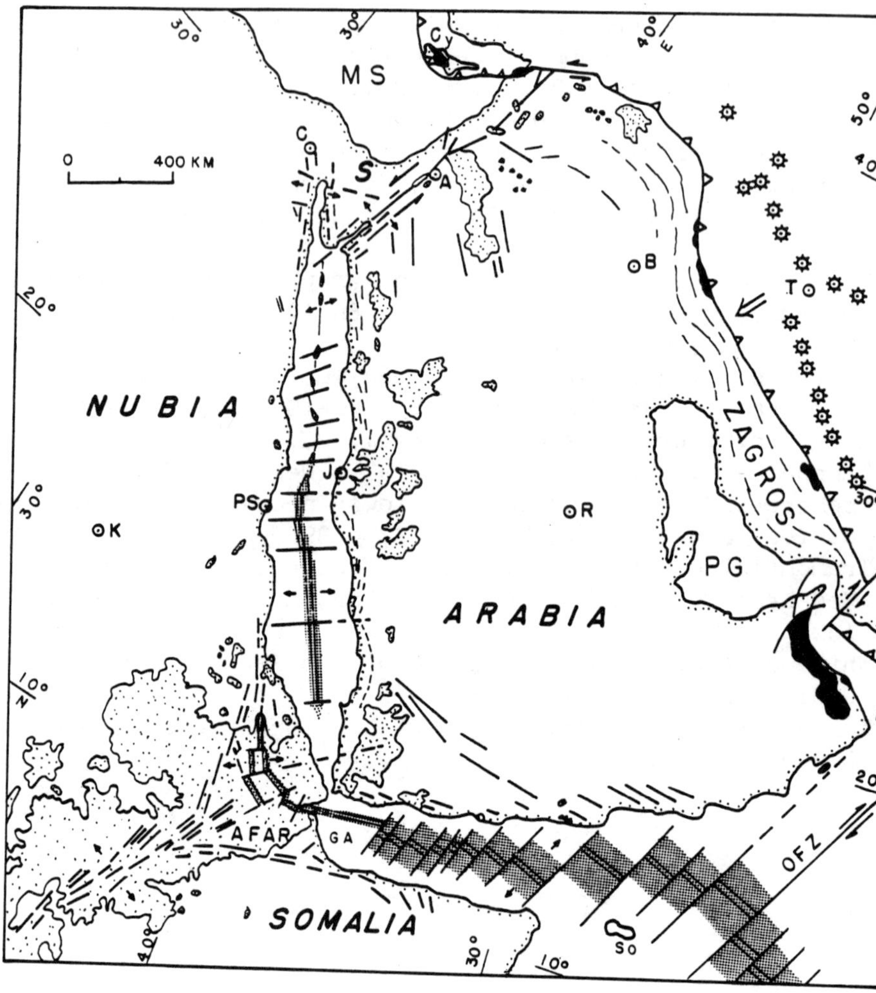

MS
Cy
0 400 KM
C
S
A
B
T
NUBIA
ZAGROS
J
PS
K
R
PG
ARABIA
AFAR
GA
OFZ
SOMALIA
SO

FIGURE 1. Plate tectonic setting for the Red Sea showing the main structural features and the distribution of Cenozoic volcanic rocks. Double lines show presently active spreading centers with the double arrows indicating spreading direction. The estimated areas of new ocean crust formed by spreading are shown heavy stipple. Volcanic fields are represented by light stipple. The asymmetry of volcanism can be easily ascertained on this map projection. Major faults are shown by heavy lines, those with barbs indicate thrusting and the direction of movement is shown by arrows. Crustal attenuation is also shown by opposite arrows in the Afar and the northern gulfs of the Red Sea. Stars indicate subduction related calc-alkaline volcanics during the late Tertiary consumption of Arabia under the Eurasian plate east of the Zagros fold belt. Short dashed lines along the Red Sea coast and in the Sinai indicate dike swarm trends. A - Amman, B - Baghdad, C - Cairo, Cy - Cyprus, GA - Gulf of Aden, J - Jeddah K - Khartoum, OFZ - Owen Fracture zone, PG - Persian Gulf, PS - Port Sudan, R - Riyad, S - Sinai, So - Socotra, T - Tehran. Two-point equidistant map projection 1:21,225,600 from National Geographic Atlas of the World, Wash. D.C., 1963.

FIGURE 2. Generalized map of the Red Sea showing the relationships between structure and volcanics. Areas of new oceanic crust are shown by a heavy stipple in the axial trough of the Red Sea. Discontinuous zones of new crust north of 22° have been established from the magnetic map of Hall et al. (1977). The oceanic crust in the Afar depression is derived from Barberi and Varet (1975). Cenozoic volcanic fields are shown in dashed stipple and associated dikes and gabbros as dotted lines or dotted clusters. Area of crustal extension shown as a very light gray screen and is estimated from magnetic patterns, dike swarms intensity and style of faulting. GE - Gulf of Elat, Z - Zabargad Island, J - Jeddah, PS - Port Sudan, TAC - Tihama Asir Complex, FI - Farasan Islands, JZ - Jizan, DI - Dahalak Islands, JT - Jubal Tair, ZI - Zubair Island, ZHI - Zukur-Hanish Islands, D - Danakil horst, A - Addis Ababa, BM - straits of Bab El Mandeb, B-I - exploratory well, M - Mansiyah-1 exploratory well.

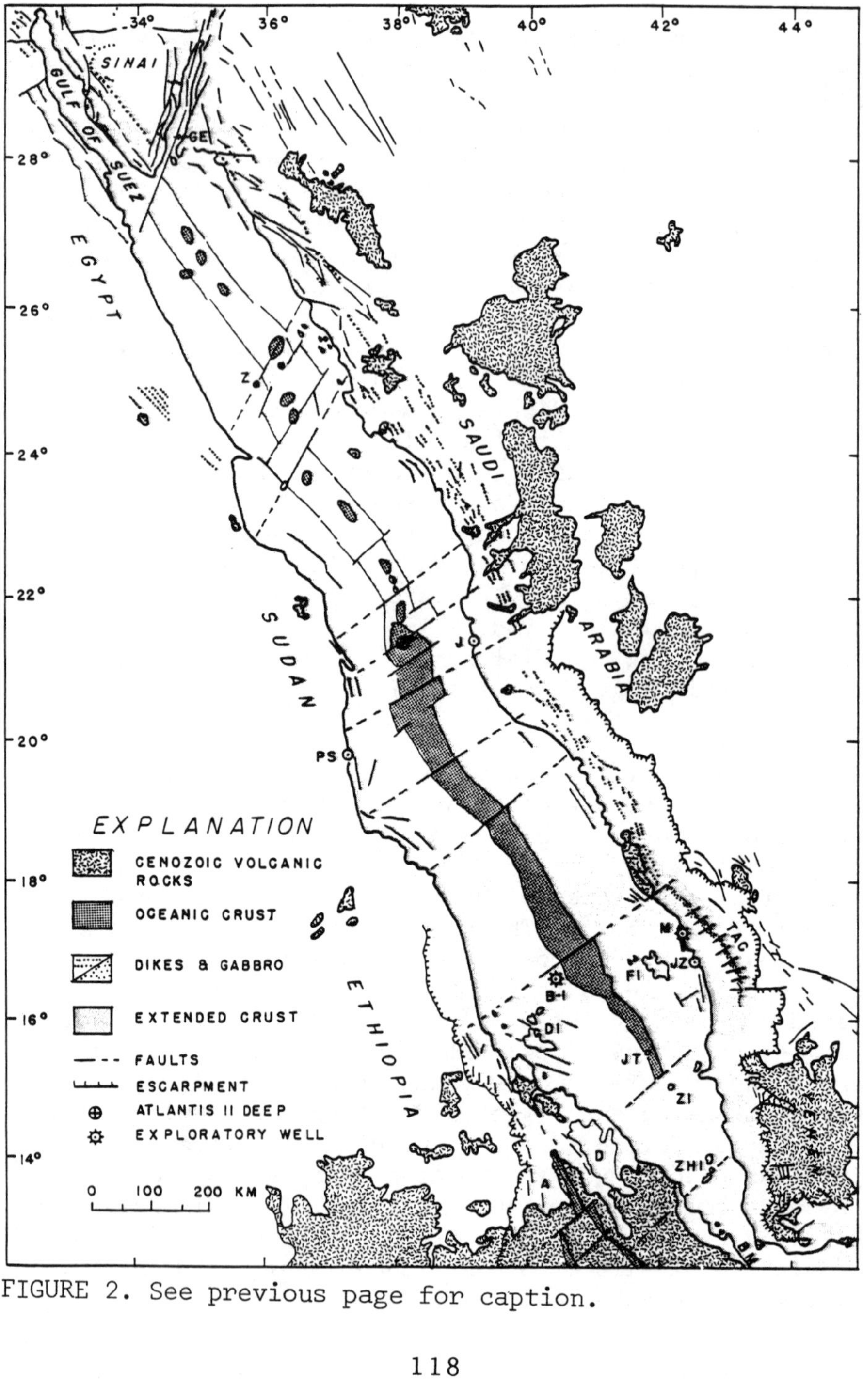

FIGURE 2. See previous page for caption.

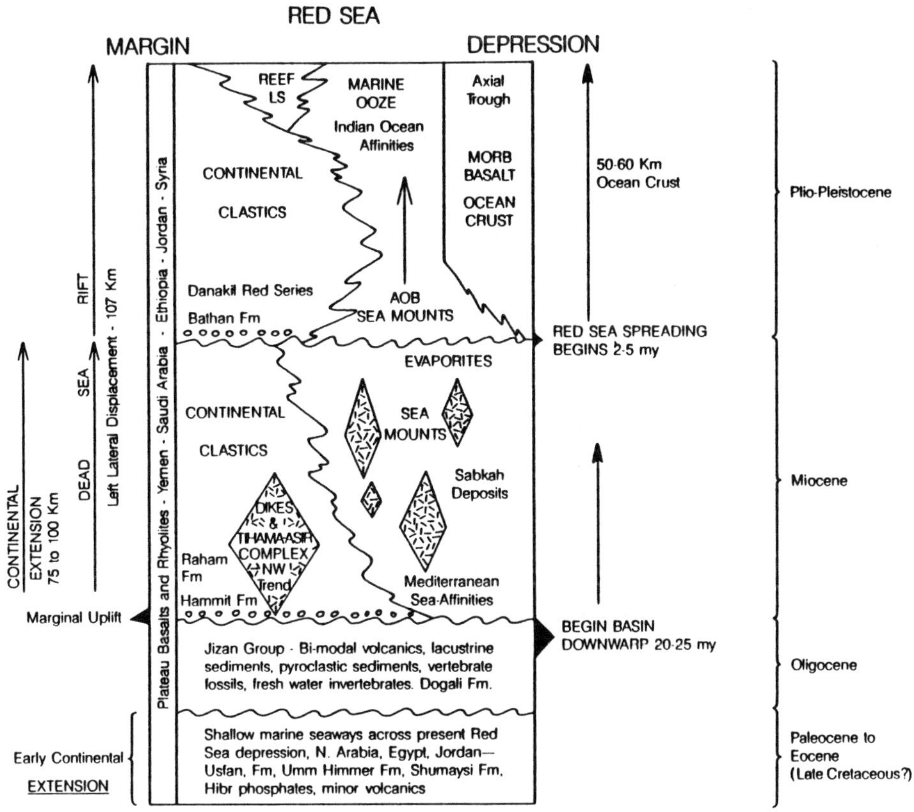

FIGURE 3. Tectonic-stratigraphic progression of the formation of the Red Sea based on the volcanic, stratigraphic structural, data from numerous geologic sources. Timing of Dead Sea rift based on K/Ar ages of dike swarms Bartov et al. (1980) and Coleman et al. (1977). Age of Jizan Group from paleontological evidence and K/Ar ages Schmidt et al. (1982). Continental extension considered to start in late Oligocene and is marked by the beginning of volcanism and block faulting. Zones of dikes and gabbros form discontinuous screens in the attenuated continental crust as typified by the Tihama Asir complex. Ocean crust formation and spreading is marked by the major unconformity above the evaporites at about 2-5 m.y. ago as mid ocean ridge basalts (MORB) form new crust in the axial trough along with alkali olivine basalts (AOB) sea mounts at the distal ends of the propagating rift.

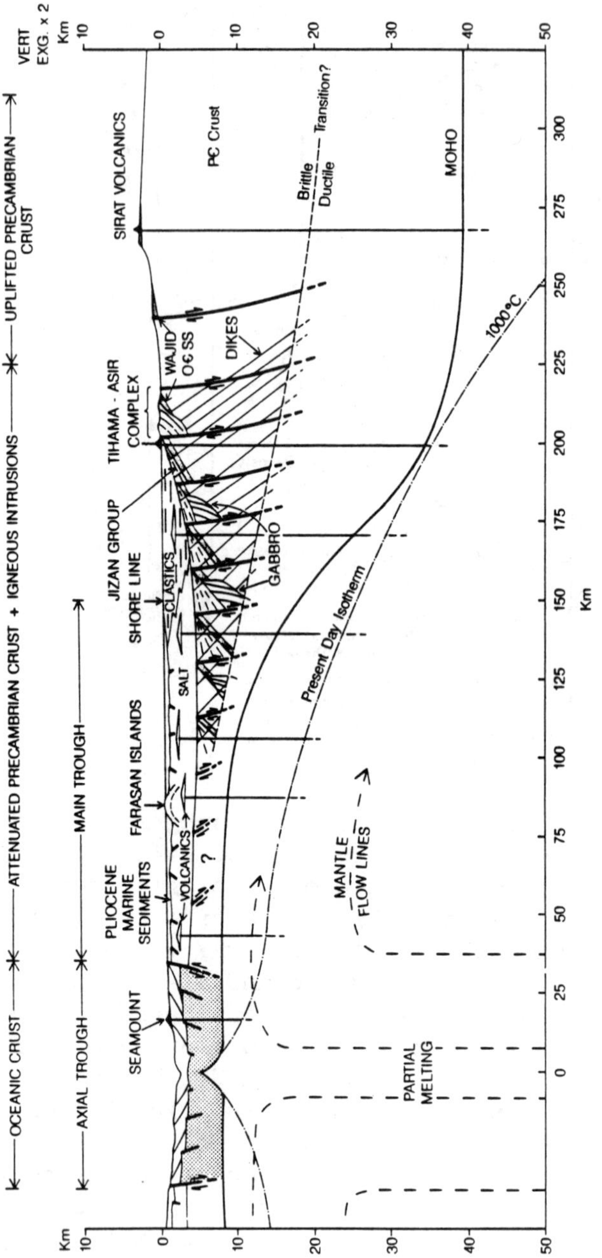

OCEANIC CRUST
ATTENUATED PRECAMBRIAN CRUST + IGNEOUS INTRUSIONS
UPLIFTED PRECAMBRIAN CRUST
AXIAL TROUGH
MAIN TROUGH
VERT EXG. x 2
Km
SEAMOUNT
PLIOCENE MARINE SEDIMENTS
VOLCANICS
FARASAN ISLANDS
SALT
SHORE LINE
CLASTICS
JIZAN GROUP
GABBRO
TIHAMA - ASIR COMPLEX
WAJID O€ SS
DIKES
SIRAT VOLCANICS
P€ Crust
Brittle
Ductile
Transition?
MOHO
1000°C
Present Day Isotherm
MANTLE FLOW LINES
PARTIAL MELTING
0
25
50
75
100
125
150
175
200
225
250
275
300
Km
10
0
10
20
30
40
50

FIGURE 4. Diagrammatic cross section of the Red Sea passive margin from the axial trough across the Farasan Island, Tihama Asir complex, and over the escarpment beyond the Sirat volcanics. Control on thickness of the crust is based on the deep seismic refraction profile in this area by Healy et al.(1982). Sedimentary thickness are estimated from stratigraphy in the Mansiyah-1 exploratory well (Gill man, 1968) and seismic refraction lines from Drake and Girdler (1964), and Tramontini and Davies (1969). Configuration of the faults is from observations at the Tihama-Asir complex and only partial flattening (lystric) is shown at the brittle-ductile transition zone. This zone is estimated to be around 500°C for typical rocks of the Saudi Arabian Precambrian shield. The 1000°C isotherm is based, in part, on the geothermal gradient determined from heat flow studies at the Mansiyah-1 well (Girdler, 1970) and the expected geotherm configuration around a slow spreading center (Tapponier and Francheteau, 1978). It is assumed that the Precambrian continental crust started to extend in late Oligocene to early Miocene contemporaneously with dike intrusions and that both processes continued up to Pliocene time when steady state spreading in the axial trough began to form MORB crust. Alkaline olivine basalts are presently erupting in small volumes along the Red Sea shoreline (Tihama Asir) and at the distal end of the propagating axial trough south of Jabal Tair (Fig.2). The diabase dikes are rotated by continued lystric style faulting and the gabbro complexes are formed along these active faults. The cone-shaped gabbro bodies develop along unconformities enlarging themselves into larger intrusions along zones of lithologic weakness in a fashion similar to that visualized along the passive margin of East Greenland (Brooks and Nielsen, 1982).

MARGINAL SEAS OF SOUTHEAST ASIA - THEIR GEOPHYSICAL CHARACTERISTICS AND STRUCTURE

DENNIS E. HAYES
Lamont-Doherty Geological Observatory of Columbia University, Palisades, New York 10964, USA

ABSTRACT

To first approximation, the petrology, seismically-determined layer structures and characteristics, magnetic properties, depths, gravimetric signatures, and heat flows of the deep Southeast Asian seas are virtually indistinguishable from the corresponding average properties of the world's ocean basins. The forces that initiate or sustain the crustal extension associated with basin formation are not the same for all of the Southeast Asian marginal sea basins. For example, the South China Sea Basin was formed by passive rifting of continental blocks. In contrast, the Parece Vela/Shikoku Basin and others were formed by "back-arc" spreading. The origin of the West Philippine Basin and of others remain unknown or controversial. The crustal structure along a transect lying near 18°N and crossing four marginal sea basins is presented and discussed in detail. The relatively small size of the Southeast Asian marginal seas gives rise to problems in dating short magnetic lineation sequences. All Southeast Asian marginal basin seafloors are Tertiary or younger and most were formed between the mid-Oligocene and mid-Miocene. These relatively short periods of crustal accretion illustrate the instability of the region. The short pulses of spreading that characterize the marginal sea basins of Southeast Asia should be manifested as recognizable tectonic events in the island arcs of the region.

Proceedings of the 27th International
Geological Congress, Volume 23, pp. 123–154
ORIGIN AND HISTORY OF MARGINAL AND INLAND SEAS

INTRODUCTION

The Southeast Asian Seas are comprised of a complex mosaic of small basins and shallow marginal seas. The basins are typically bounded by conspicuous morpho-tectonic features such as island arcs, deep-sea trenches, and aseismic ridges. The general geological history of many of the southeast Asian marginal seas is now known and reveals that there are striking similarities in the timing, duration, and nature of the formation of these basins. There appear to be at least three categories of basin evolution that characterize the development of the region as a whole.

Two important problems remain to be resolved before we can hope to understand the relationships between the evolution of the marginal seas and the surrounding land areas in the context of small-scale plate tectonics. The first problem involves our inability to determine if the differences in the observed geophysical characteristics between the Southeast Asian marginal seas are the result of fundamentally different processes of basin formation, scale differences, or varying responses unique to a particular geographic/geologic locale. The evolution of the region no doubt reflects a complex matrix of time, space, and process, the effects of which are barely known, let alone separable! The second problem involves our inability to determine the ages or to recognize the tectonic fabrics and gross geophysical characteristics of portions of the major marginal sea basins and of most of the smaller basins. Hayes (1982) has pointed out that many marginal sea basins appear to be composed of two or more sub-basins, each with unique characteristics. The horizontal scale of the individual tectonic units (platelets) ranges from about 1000 km down to less than 100 km. The small size of these units has pre-empted many of our standard analysis techniques, most notably the mapping and identification of characteristic marine magnetic lineations. Dating the crust in conjunction with recognizing the orientation of the crustal fabric are crucial prerequisites for understanding the plate kinematics of this complex region and for examining the onland geological responses to plate interactions at collisional boundaries.

The petrology, seismically-determined crustal layer structure and characteristics, magnetic properties, depths, gravimetric signatures, and heat flow of the southeast marginal seas are virtually indistinguishable from the corresponding average properties of the world's major oceans' basins. The variations in these properties observed within southeast Asian seas are about the same as the variations observed within a single major ocean basin. Hence, the principal characteristic that sets marginal seas apart from the major ocean basins is their small size.

Two recent review papers discuss the evolution and characteristics of marginal basins as a whole and establish a useful global perspective for discussing the marginal seas of southeast Asia (Weissel, 1981; Taylor & Karner, 1983). The theme of both of these review papers is the categorization of the world's marginal basins into: 1) those that have evolved through back-arc spreading of the type defined by Karig (1974); 2) those that demonstratively have not evolved through this mechanism; and 3) those of unknown or uncertain evolution.

The concept of "back-arc basin formation" implies a genetic relationship whereby the basin is formed by extension along the volcanic arc side of an active trench-arc system. A specific timing of events and a definite relationship between the orientation of the volcanic arc and the tectonic fabric of the associated back-arc basin is explicit in the Karig (1974) model. The Mariana Trough and Parece Vela Basin are both considered prototypes of active and inactive back-arc basins, respectively. However, the mere existence of an active island-arc trench system bordering a marginal basin is not itself sufficient to insure that a back-arc basin exists or will develop.

An example of a non back-arc basin is the South China Sea Basin which has evolved through normal rifting and spreading processes. The South China Sea Basin is bounded on the north and south by passive continental margins, and the evolution of the South China Sea Basin can be likened to that of a small Atlantic-type ocean basin (Taylor and Hayes, 1980 and 1983). The presence of the active Philippine Island arc system which bounds the South China Sea on the east has been essentially unrelated to the evolution of the South China Sea Basin. The Tasman Sea (Hayes and Ringis, 1973) is another non-back-arc basin, which has evolved in a fashion similar to the South China Sea. The Aleutian Basin is composed of a piece of old, trapped Mesozoic oceanic crust which has been isolated by the development of a much younger arc-trench system to the south (Cooper et al., 1976). The Aleutian Basin also is a non-back-arc basin even though it resides in a back-arc position.

The third category of marginal basins involves basins of unknown or uncertain origin. Their uncertain evolution is primarily the consequence of inadequate information regarding their age or tectonic fabric or an ambiguity in the genetic relationship of these basins to other linear tectonic features (island arcs, passive margins, aseismic ridges).

Examples of uncertain basins are the Celebes Sea Basin, the Sulu Sea Basin, and the West Philippine Basin. Figure 1 shows nine principal marginal basins in southeast Asia. These are: the Caroline Basin, the Mariana Trough, the Parece Vela Basin, the Shikoku Basin, the Bonin Trough, the West Philippine Basin, the South China Sea Basin, the Sulu Basin, and the Celebes Basin. In addition, there are other smaller features which no doubt represent incipient or very young basin formation. These include the Okinawa Trough, the Ayu Trough, and perhaps the Sorol Trough. Many of the major basins appear to be characterized by one or more sub-basins with some properties (e.g., tectonic fabric, magnetic lineation fabric, or other geophysical characteristics) that are noticeably different from those of the main basin (see Figure 1).

As previously mentioned, determining the age and the age gradients of the crust of the Southeast Asian marginal seas is crucial to unraveling the

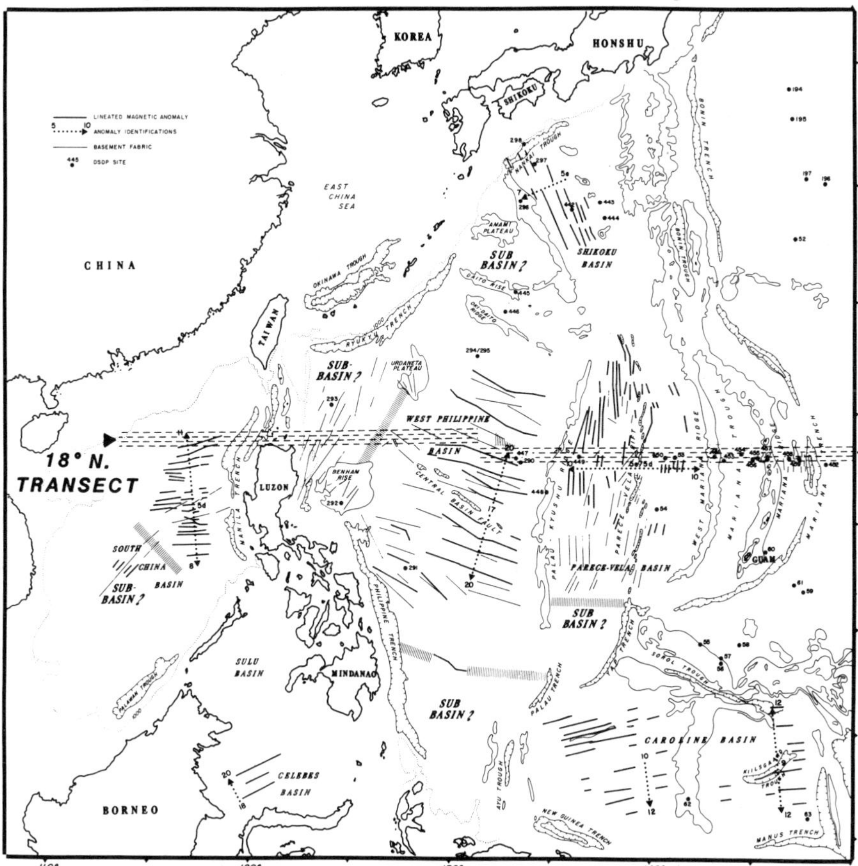

FIGURE 1. Tectonic and magnetic lineaments in Central Southeast Asian seas. Magnetic identifications are from sources referenced in the text. Note the locations of the five sub-basins that have been tentatively identified. The location of the 18°N Transect Zone Section shown in Figure 2 is indicated.

tectonic evolution of the region. Such age dating can only be accomplished by direct sampling through drilling and dredging or by identifying dateable marine magnetic lineations. As shown in Figure 1, the total number of DSDP holes in the southeast Asian region is less than 30 and these sites are contained within just four of the basins. Many of these holes did not penetrate to basement, and thus they provide only limited, minimum crustal age data. Extensive dredging has been carried out in many of the basins, but only a handful of the samples have been fresh enough to allow reliable radiometric dating and few such dates exist in the literature. Thus, most Southeast Asian basins have no directly determined age dates and many remain virtually unsampled.

The analysis of marine magnetic lineations has proved to be the most successful approach to dating the basins but this method has also met with important limitations in Southeast Asian seas. The small sizes of the marginal sea basins compromise one's ability to recognize a unique sequential pattern of magnetic lineations which is a prerequisite for magnetic anomaly dating of the underlying oceanic crust. In Southeast Asia, the problem of mapping magnetic lineations is complicated further by the fact that most of the basins presently reside or were formed in low geomagnetic latitudes and several exhibit northward striking magnetic and tectonic fabrics. Hence, even those magnetic lineations which are present are often recorded with very low amplitudes and with questionable fidelity. Despite these difficulties, marine magnetic lineations have now been mapped and dated for major portions of the Parece Vela/Shikoku Basin system, the West Philippine Basin, the South China Sea Basin, and the Caroline Basin. These magnetic lineations are shown in Figure 1. Unfortunately, major portions of the region remain undated by this or any other method and the need for several well-placed DSDP holes is painfully apparent.

R E C E   V E L A
PARECE
RI
24
30
48
6.8
33
41
30
42
49
37
51
70
82
77
85
25V21TS
39V35
193C20
192C20
23V21TS

Basin parallel to the tectonic fabric. Hence, the results presented along this section for the two western basins cannot be extrapolated in a north-south direction.

The Pacific Plate

The Pacific plate crust has variable thickness ranging from 5.5 km to 7 km. It is defined by typical layer 2 and layer 3 velocities, but the combined thickness of these two layers appears to be slightly thinner than the Pacific basin averages (LaTraille and Hussong, 1980). There also appears to be an unusually thick layer with velocities characteristic of layer 2A (3.5 to 4.5 km/s). Because layer 2A is normally not found in very old oceanic crust (Houtz and Ewing, 1976), LaTraille and Hussong (1980) have suggested this layer may be the result of extensive mid-plate volcanism. Alternatively, this layer may represent very thick chert layers (Hussong and Uyeda, 1981). DSDP Site 452 terminated after only 50 meters of drilling in Neogene sediments in impenetrable cherts of probable Campanian age (Hussong, Uyeda et al., 1981). Thus, *Glomar Challenger* Leg 60 scientists argued that any Paleogene sediments sampled on the arc side of the trench are not likely to have been derived from the Pacific plate.

The Mariana Trench

At this locale the Mariana Trench is about 8.6 km deep and is nearly sediment free. Two DSDP sites, 460 and 461, were drilled on a small terrace of the inner-trench wall. These combined sites had less than 150 meters penetration, and this with incomplete core recovery. Site 460, however, sampled relatively undisturbed sediments that record a gradual subsidence from above the CCD to a present depth of approximately 6500 meters. All existing data point to the absence of any significant accretionary prism at this locale and suggest that subsidence accompanied by tectonic erosion is the dominant operative process along the Mariana Trench.

There is no deep crustal information from the zone directly beneath the trench axis.

The trench is associated with a narrow free-air gravity anomaly (FAA) minimum reaching values of about -220 mgal. Small amplitude magnetic anomalies (peak to peak ∿200 gammas) characterize the Pacific plate up to the trench axis; westward magnetic anomaly amplitudes are even smaller over the forearc region. The small amplitude anomalies on the Pacific plate appear to be lineated over significant distances (Hussong, 1981) and although presently undated, probably represent Early to Middle Jurassic age crust.

The Mariana Forearc

The crustal thickness in the Mariana forearc appears to range from approximately 4.5 km in the east to 9+ km in the west. Forearc sediments are also thickest near the volcanically active Mariana Ridge. The general velocity structure of the crust underlying the forearc is not dramatically different from that associated with normal oceanic crust. The eastern side of the forearc appears to be subsiding as evidenced by numerous high-angle normal faults striking subparallel to the trench and arc. The western forearc is also disrupted by normal faults but these may be the consequence of uplift of the volcanic arc itself (Mrozowski and Hayes, 1980).

DSDP Sites 458 and 459 were drilled in the forearc region. Igneous basement was penetrated at both sites; each site exhibited local sediment reworking and redeposition throughout the drilled section, consistent with the idea of extensive faulting throughout the forearc. The sedimentary record at the two sites is quite different and must reflect the effects of local tectonism. The deep basement at site 458 consists of andesites interbedded with basalts which are chemically similar to the island arc tholeiitic basalts that dominate the basement at site 459 (Hussong & Uyeda, 1981). The shallow basement at site 458 consists of a variety of petrologic types, including boninites (Meijer et al. 1981). Heat flow measured at both sites 458 and 459 is low, below 40 mW/m^2 (Uyeda and Horai, 1981), and typical of most forearc regions.

The Mariana Arc

DSDP site 457 was drilled on the flank of the active volcanic arc but did not penetrate through the recent volcaniclastic debris that blankets the area. The volcanic arc is associated with a high-amplitude, short wavelength magnetic anomaly and with free air gravity anomalies in excess of +100 mgals. The gravity data suggest that the Mariana Ridge is at least partially compensated by a deep crustal root although there is no good refraction data from beneath the arc to determine the exact crustal thickness there.

The Mariana Trough

All available geophysical data indicate that the Mariana Trough is the site of active seafloor spreading. This area has become the prototype example of back-arc marginal basin development. The work of Karig (1978), Anderson (1975), Bibee et al. (1980), Hussong and Sinton (1983), and of others show that the Mariana Trough exhibits the ridge-transform fault offset geometry typical of mid-ocean ridges. Bibee et al. (1980) made tentative correlations of the magnetic lineations within the trough and suggested half-spreading rates of $\sim$1.5 cm/yr. Their anomaly correlations could not be extended beyond 50 km from the center of the trough. Hussong and Fryer (1981) using more extensive data, argued that the trough is opening at a half-rate of 1.65 cm/yr.

The Mariana Trough crust is composed of an unusually thick layer 2A with velocities of about 3.2-4.3 km/s. The crust is approximately 5 km thick, atypically thin for shallow seafloor, but consistent with the mean positive gravity anomaly across the entire trough of $\sim$+50 mgal. Except for varying thicknesses of layer 2A and 2B, the thickness and velocity of layer 3 and depth to Moho is more similar to that of older oceanic crust than to very young oceanic crust. Four DSDP sites were drilled in the region: DSDP site 453 at the extreme western edge of the Trough, site 454 near the center of the trough, and sites 455 and 456 on the western flank of the active volcanic arc. Relatively little sediment exists in the central portion of the Mariana Trough. At site 453, both paleontologically and magnetically determined ages of the basal sediments suggest an age of the underlying crust of approximately 5+ million years and indicate that the average half-spreading rate for the trough has been ∿2.2 cm/yr (Hussong & Uyeda, 1981).

The igneous basement rocks recovered at sites 454 and 456 establish the likelihood that fundamentally different magma types were supplied simultaneously near the center of the Mariana Trough. The heat flow values obtained both from the DSDP sites (Uyeda and Horai, 1981) and from surface-ship measurements (Hobart et al., 1979; Anderson, 1975) indicate that heat flow values in the trough are extremely variable and include some very high values. Hydrothermal circulation within the upper crustal layers is clearly an important process in this region.

The West Mariana Ridge

Site 451 was drilled near the eastern side of the West Mariana Ridge, the youngest of the remanent arcs in this region. Most of the 925m of sediment sampled from hole 451 consisted of upper Miocene volcaniclastic breccias, conglomerates, and tuffs (Scott et al., 1980). Pillow basalt fragments were recovered from the base of the sediments. Scott et al. (1980) identify three phases of volcanic activity resulting in the gradual buildup of over 1 km of volcaniclastic debris on top of oceanic crust. This interpretation is consistent with the limited seismic data reported by Langseth and Mrozowski (1980) as part of the pre-drilling site survey. Scott et al. (1980) believe volcanic activity ceased about 5 million years ago and that the area received only biogeneous sediments since that time and has subsided slightly. A substantial crustal root, extending to perhaps 13 km and resulting in a 10+ km crustal thickness, is inferred from the observed free-air gravity anomaly, the topographic relief, and isostatic considerations.

The Parece Vela Basin

This basin has been studied extensively by Mrozowski and Hayes (1979) who show, by the identification of north-south trending magnetic lineations, that the basin evolved through seafloor spreading processes during early Oligocene to middle Miocene time. The Parece Vela Basin is dominated by a north-south tectonic fabric, including a spectacular rift with relief in excess of 3 km. The rift is often found near the axis of symmetry of magnetic lineations within the center of the basin. The morphology and shallow crustal structure of the eastern and western flanks of the Parece Vela Basin is dramatically different. The eastern flank exhibits a smooth morphology, formed by the blanketing of the basement relief by thick volcaniclastic sediments derived from the West Mariana Ridge. Sediments range in thickness from 3+ km adjacent to the ridge and gradually thin to a few hundred meters near the Parece Vela Rift. In marked contrast, the relief of the western flank is very rugged with only thin (less than 100 meters thick) pelagic sediment cover present over most of the region.

Both flanks of the Parece Vela Basin are associated with the presence of thick layer 2A overlying a typical layer 2, in turn overlying normal layer 3 velocities and Moho. The total crustal thicknesses throughout the basin ranges between 5-6 km. Although there is no seismic refraction or sonobuoy data from near the Parece Vela rift, Figure 2 illustrates that unless there are major fluctuations in depth of the layer 2-3 boundary, the Parece Vela Rift may reach downward, close to the layer 2-3 boundary. Small ponds of sediment are found in the deep Parece Vela Rift and may define unique drilling targets for those interested in sampling layer 3 in relatively young oceanic crust.

Magnetic anomalies within the Parece Vela Basin are poorly developed. Mrozowski and Hayes (1979) have identified anomalies 5E through 10 over the western flank of the Parece Vela Basin. Anomaly amplitudes are severely attenuated over the eastern flank, a pattern very similar to that found to the north in the Shikoku Basin. Magnetic lineations are typically less than 200 gammas peak to peak amplitude and cannot be traced northward beyond about 23°N nor southward beyond about 14°S. Spreading in the Parece Vela Basin began approximately 30 million years ago. In contrast, the initiation of spreading in the Shikoku Basin began about 25 million years ago. This implies that a major shear boundary existed between the two basins between 30 and 25 m.y.B.P. Such a shear boundary has not yet been recognized.

The Parece Vela Rift is associated with a relative free-air gravity <u>minimum</u> of about 100 mgal, but that can largely be explained by the uncompensated relief of the Parece Vela Rift. The average free air gravity value across the Parece Vela Basin is approximately +15 mgal and represents a nearly constant "step-down" of 35 mgal from the average FAA value of the Mariana Trough.

The western flank of the Parece Vela Basin is wider by more than 100 km than the eastern flank across this profile. Either asymmetric crustal accretion has occurred or a significant, discrete ridge-crest jump(s) occurred, probably during the early stages of spreading. The western flank of the Parece Vela Basin is extremely rough with numerous deep rifts which could represent previous spreading centers or ridge-crest jumps. One such rift is obvious in Figure 2 near the location of DSDP site 449.

Two DSDP sites were drilled in the Parece Vela Basin. Site 450 on the eastern flank is near the center of the basin and site 449 is on the western flank near to the Kyushu-Palau Ridge. Typical MOR basalts were recovered at site 449 and the paleontological age of the basal sediments was close to that predicted from the magnetic lineations. At site 450, chemically similar tholeiitic basalt was sampled, but its age, trace-element chemistry, and the character of the sediment basalt contact led Scott et al. (1980) to conclude that this basalt resulted from off-axis intrusive activity which occurred several million years after the creation of the underlying oceanic crust.

Scott et al. (1980) outline the likely history of the region roughly as follows: Oceanic crust at site 449, formed about 24-26 m.y. ago, was the site of pelagic sedimentation, initially above the CCD but subsided below it after only 3-5 m.y. Influx of disseminated ash and other volcanic debris between approximately 20 m.y. ago and 13 m.y. ago is interpreted as marking the period of pyroclastic volcanic activity on the West Mariana Arc. To the east, site 450 was also a site of a rapid influx of volcanigenic turbidites derived from the west Mariana Ridge but was isolated from this between 9 and 5 m.y. ago. After this time, probably as a consequence of crustal cooling and faulting within the basement, the region subsided below the CCD.

The Kyushu-Palau Ridge

DSDP site 448 was drilled on the eastern flank of the Kyushu-Palau Ridge. The ridge itself is underlain by a thick sequence of low velocity (3.5 km/s) material which consists of flows, sills, dikes, and volcaniclastic breccias which were built on crust similar to layer 2 but probably thickened by arc-related intrusives (Scott et al., 1980). This layer, in turn, rests on 6.6 km/s layer 3, and overlies an 8.0 km/s Moho. The drilling results are consistent with a history which involves arc-related volcanism commencing in the Late Eocene and persisting into the the Mid-early Miocene. Volcanic activity ceased, roughly at the time of inter-arc rifting and the initial stages of formation of the Parece Vela Basin. The crustal thickness beneath the crest of the Kyushu-Palau Ridge is about 9 km. There is virtually no gravimetric expression of the Kyushu-Palau Ridge.

The West Philippine Basin

Just west of the Kyushu-Palau Ridge is a thick wedge of presumably volcaniclastic sediments similar in disposition to those along the west flank of the West Mariana Ridge; they have not yet been sampled. The West Philippine Basin lies to the west of the Kyushu-Palau Ridge and is bounded on the west by the complex Philippine island arc system and its northward extension. The West Philippine Basin represents the largest of the marginal basin systems of Southeast Asia and perhaps the one whose origin is most controversial (see, e.g.. Uyeda and Ben Avraham, 1972; Watts and others, 1977; Louden, 1976; Shih, 1980; Mrozowski and others, 1982; and Hilde and Lee, in press). Most of these investigators identify a magnetic lineation fabric throughout the region that trends in a WNW to NW direction. The Central Basin "Fault" is near to or coincident with the axis of magnetic lineation symmetry. The magnetic lineations can be traced only within the eastern half of the West Philippine Basin. Magnetic lineation identifications of Watts et al. (1977) are consistent with the limited crustal age information derived from DSDP sites in this basin. DSDP site 292 on the Benham Rise and site 293 lying in the extreme northwest of the basin provide little information pertinent to the evolution of the principal portion of the West Philippine Basin. Because there is relatively poor age gradient information from the DSDP sites, various different interpretations of the magnetic lineation ages are possible that do not conflict with the DSDP results. The differences between the various interpretations (e.g., Watts and others, 1977; Shih, 1980; and Hilde and Lee, in press) primarily involve differences in the estimated rates of crustal accretion and the time when spreading activity within the basin ceased.

The composite trend of the Central Basin Fault is approximately 300° while the trend of the magnetic lineation azimuth is nearer 285° to 290°. Lewis and Hayes (1980) have shown that at least portions of the Central Basin Fault represent individual crustal segments trending nearly 285° and offset en echelon along numerous, closely-spaced faults.

Most investigators basically agree that the principal part of the basin was formed during the Eocene and differ only in whether or not spreading continued into the Early Oligocene. Resolution of these uncertainties awaits additional sampling through deep drilling, designed specifically to address this question and coupled with regional multibeam bathymetric (tectonic fabric) mapping. A most important point is that <u>both</u> the magnetic lineation fabric and the tectonic fabric trend at very high angles to the bounding remanent Palau-Kyushu arc and to the active Philippine arc system to the west. It is clear, therefore, that the West Philippine Basin did <u>not</u> evolve through back-arc spreading in association with an active Kyushu-Palau Arc. The possible back-arc relationship of the basin to older volcanic terranes of the Philippine island-arc complex has been discussed by Mrozowski et al. (1982).

The total crustal thickness throughout the West Philippine Basin appears to be slightly thinner than the average of oceanic crust. Throughout the region, layer 2, associated with 5.0 km/s velocity, is overlain by a 6.6 km/s layer, which in turn overlays Moho, whose velocity along the transect appears highly variable, ranging from 7.6 to 8.6 km/s. In the eastern half of the basin and to the north of the Central Basin Fault, velocities associated with layer 3 have been found at anomalously shallow depths below the seafloor. Several independent measurements (see Fig. 2) have determined that layer 2 is slightly less than 1 km thick within this region. This area was one of the initial drilling targets for DSDP leg 59 but was abandoned because of operational limits of the <u>Glomar Challenger</u>.

Except for local gravity anomalies associated with the high-amplitude topographic relief of the Central Basin Fault and the north-trending ridge structures along the eastern Luzon margin, free-air gravity anomalies in the West Philippine Basin average -5 mgals. The entire West Philippine Basin is unusually deep for oceanic crust of Eocene age but the gravity data indicates near perfect isostatic compensation and is consistent with the relatively thin total crustal thickness defined by the rather sparse seismic refraction control.

Because the Central Basin Fault trends only slightly north of west, in Figure 2 the Central Basin Fault appears to be displaced toward the western side of the basin, but this is an artifact of the east-west projection. Magnetic anomalies across the transect appear to be of relatively small amplitude, but this is also the result of this particular section being oriented nearly parallel to the magnetic lineation trends. An area exists between the North Luzon Ridge and the Central Basin Fault where both the tectonic fabric and the magnetic fabric appear to be trending more nearly north-south than east-west. On the basis of this information, it has been hypothesized that west of a line extending between the Benham Rise and Urdaneta Plateau, a separate sub-basin of the West Philippine Basin exists (Hayes, 1983; Mrozowski et al., 1982).

The East Luzon Margin

At latitude ∿18°N there is no direct evidence of active subduction along the east flank of the North Luzon Ridge although Lewis and Hayes (1983) have identified tectonic features and geophysical characteristics associated with subduction (including major thrust faulting) only 200 km south of the 18°N transect zone.

The East Luzon Trough is believed to represent the site of a former trench that was active during Oligocene time along eastern Luzon (Lewis & Hayes, 1983). The remainder of the subduction complex lies to the west of the trough along the eastern Luzon margin and the Sierra Madre mountains. Because there is no strong evidence of active subduction along the North Luzon Ridge, the relative convergence between the Philippine Sea plate and the Luzon Arc is probably accomodated by a complex pattern of wrench faulting and crustal shortening across the

North Luzon Ridge region (Lewis and Hayes, in prep.). To the south, plate convergence is taken up through normal subduction tectonics along the Luzon Trough and the Philippine Trench (see Lewis and Hayes, 1983). The deep crustal structure beneath the North Luzon Ridge is unknown. Seismicity throughout the region is diffuse (Hamburger et al., 1983). At 18°N the Cagayan Valley Basin of northern Luzon extends northward offshore over the western half of the North Luzon Ridge complex. The offshore portion of the Cagayan Valley Basin contains relatively flat-lying sediments up to 2 km in thickness.

The West Luzon Margin

The Manila Trench subduction zone and the associated Luzon Trough forearc basins have been studied in some detail (Ludwig, 1970, and Hayes & Lewis, in press, Lewis & Hayes, in press). The forearc region includes a large accretionary prism composed of several kilometers of relatively low-velocity material($<$3.6 km/s). The North Luzon Trough forearc basin contains 2+ km of relatively flat-lying, turbidite sediments derived from the north and east. A diffuse zone of seismicity is present and Hamburger et al. (1983) show a poorly defined eastward-dipping Benioff zone extending to depths of only about 100 km. The active volcanic arc associated with this subduction is located along the east side of the North Luzon Ridge and includes the Batan and Babayan Islands. Neither the time of initiation of subduction along the Manila Trench nor the past or current rates of plate convergence are well known. Hayes and Lewis (1983) present loosely-constrained arguments that subduction began in Late Oligocene or Early Miocene time and that _average_ rates of crustal convergence have been slow--about 1 cm/yr.

The South China Sea Basin

To the west of the Manila Trench lies the deep South China Sea Basin. The crustal structure of the South China Sea Basin is relatively well known (Ludwig et al., 1979; Taylor & Hayes, 1980; Taylor &

Hayes, 1983). Average crustal thicknesses throughout the northern part of the basin are 6+ km. This slightly greater than normal crustal thickness can be accounted for by the unusually thick sediments which, in many places, exceed 2 km. The crystalline portion of the oceanic crust is seismically indistinguishable from that of average oceanic crust, although there is considerable variation throughout the basin regarding the apparent thickness of layer 2. Layer 3 velocities are typically 6.7 km/s and measured Moho velocities range from 8.0 to 8.4 km/s.

The South China Sea Basin was formed during Early Oligocene to Middle Miocene time (Taylor & Hayes, 1980, 1983). This conclusion is based primarily upon magnetic lineations that are well mapped throughout the eastern half of the South China Sea basin. Portions of the continental Asian mainland were rifted and carried southward by seafloor spreading. The South China Sea Basin was formed as a result and records an east-west oriented magnetic and tectonic fabric. Spreading ceased about 17 m.y. ago, approximately coincident with the time that the rifted Asian continental blocks collided with portions of Palawan, Mindoro, and Borneo.

It is important, again, to note that the predominant tectonic fabric of the South China Sea Basin lies at a very high angle to the north-trending Philippine arc complex. Both the northern margin (the China continental margin) and the southern margin (Reed Bank/dangerous-ground region) of the South China Sea are passive continental margins.

The relatively deep portion of the basin lying to the southwest (see Figure 1) is characterized by a different magnetic and tectonic fabric from the area to the north and east. Though existing data are inadequate to define the detailed characteristics there, all evidence points to a somewhat delayed initiation of spreading in the southwest sub-basin along a NE-SW trending zone of accretion. The earliest phases of spreading in the eastern sub-basin must have been accomodated in the west by extensive stretching of continental crust. Such stretching, with its associated subsidence and thermal effects could have important implications regarding the relative petroleum potential of shelf basins along these margins (see Hayes, 1983, and in press).

The magnetics profile across the South China Sea Basin shown at the top of Figure 2 is nearly parallel to the strike of the magnetic lineations and specific magnetic anomalies are therefore not obvious. Heat flow values in the South China Basin range from about 90 mW/m^2 to 110 mW/m^2 and vary in a fashion consistent with predictions obtained from the empirical heat flow versus age relation of Parsons & Sclater (1977) (see Fig. 8 from Taylor and Hayes, 1983).

The evolution of the South China Sea Basin is somewhat analogous to that of the Atlantic, except that spreading and continental drift took place on a much smaller scale and occurred over a brief interval of time during the middle Tertiary. To date, there have been no drill holes sited in the South China Sea Basin, although the area provides an ideal natural laboratory for testing various models of crustal extension and subsidence in passive continental margin environments.

Transect Summary

The 18°N transect nicely illustrates the three major types of marginal seas that are found in Southeast Asia. There are basins which seem clearly to have resulted from back-arc spreading, specifically the active Mariana Trough and the now inactive Parece Vela Basin. The West Philippine Sea Basin is a large marginal sea whose origin remains controversial. It appears to have evolved through relatively normal seafloor spreading processes and has characteristics that are typical of oceanic crust in the major ocean basins. Its relationship to major tectonic lineaments, namely the Kyushu-Palau Ridge and the North Luzon Ridge, remains enigmatic. Its genesis in relation to other hypothetical arc-ridge segments such as found in the east central Philippines or perhaps the Diato Ridge complex remains speculative at best.

The Philippine Island arc complex appears to represent an excellent example of various periods of plate collision and coalescence of island arc terranes of various ages. To the south of our transect line through central Luzon, active subduction is taking place on both sides of Luzon.

The South China Sea Basin has evolved through rifting of passive continental margins and is a non-back-arc basin. That its eastern boundary coincides with an active trench system is incidental to its evolution.

The geophysical characteristics described for each of the basins along this transect have not been mapped over the entire morpho-tectonic extent of the individual basins. The South China Sea Basin, the Philippine Basin, and the Parece Vela Basin, all contain regions that are represented by geophysical characteristics which set them apart as unique sub-basins. The tectonic histories of these sub-basins vary in some important respect from those of the principal portion of the basins.

Of the other basins that comprise the southeast Asian marginal seas but which have not been discussed in detail, the Bonin Trough can be considered an analogue of the Mariana Trough. The Shikoku Basin is a slightly younger analogue of the Parece Vela Basin to the south. They probably represent one basin system throughout most of their spreading histories. The Okinawa Trough lying to the north of the Ryukyu Trench can be considered a slightly younger and sediment filled version of the Mariana Trough. Too little is known of the geophysical characteristic of the Celebes Basin or Sulu Basin to speculate on their relationship to any of the evolutionary models discussed in the context of the transect. The only other major basin in the general region is the West Caroline Basin, the evolution of which has been discussed in considerable detail (see e.g., Hegarty et al., 1983, and Hegarty and Weissel, in prep.).

Much has been speculated regarding the distribution of crustal depths as a function of age for marginal sea basins. It has generally been recognized that the crustal depths of the Southeast Asian basins consistently lie below those depths predicted by the empirical standard for the North Pacific and North Atlantic oceans developed by Parsons & Sclater (1977). Hayes and Caress, in prep., have re-examined the age-depth data from all marginal sea basins and have concluded that as a group they are indistinguishable from the linear age-depth global relationship defined by Hayes (1983). However, if the marginal sea basins of Southeast Asia are considered as an independent sample, they define a depth vs. square root of age linear trend with a gradient that is considerably steeper than that of either the Parsons & Sclater (1977) or Hayes (1983) standards (see Fig. 3). There is nothing, however, to preclude

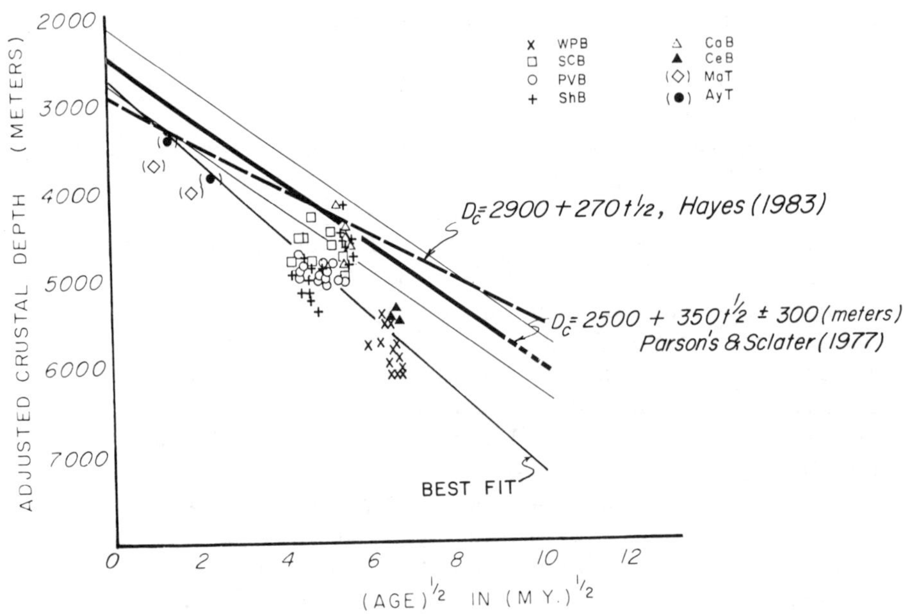

FIGURE 3. Adjusted crustal depths of selected Southeast Asian seas plotted against the square root of crustal ages (inferred). All values represent 1° x 1° averages as determined from averaged depth,

sediment thickness, and age data. All basins except the Mariana Trough and Ayu Trough have been dated by identifying the magnetic lineations mapped within the basins. The Mariana Trough and Ayu Trough data points are derived, respectively, using the age estimates of Hussong and Uyeda (1981) and Weissel and Anderson (1978). The global empirical standard for major oceans proposed by Parsons and Sclater (1977) for crust $\leq$70 m.y. old is given by the straight line equation: $D = 2500 + 350\ t^{1/2}$ where t = age and D = adjusted crustal depths. The $\pm$ meters represents the estimated standard deviations of the Parsons and Sclater (1977) data set. Note that although some values from the Shikoku, Caroline and South China Sea basins fall within the predicted depth range, the vast majority of values lie systematically 600-1000 m below the Parsons and Sclater curve. The global depth-age standard proposed by Hayes (1983) is shown for reference. The least squares linear "best-fit" from Hayes and Caress (in prep.) is also indicated. This figure is modified from Hayes (1982).

the likelihood that, to first order, the crustal depths in the marginal sea basins are controlled primarily by cooling, contraction, and subsidence in association with seafloor spreading processes.

DISCUSSION AND CONCLUSIONS

The geophysical characteristics of the major portions of many of the marginal basins in Southeast Asia are now relatively well known. The single most important unresolved problem stems from our inability to determine or even reasonably bracket the rates and directions of plate convergence across the complex multi-plate boundaries that characterize the Southeast Asian seas region. Hayes (1982) has pointed out that the major pulses of seafloor generation which have occurred throughout the area are short in duration and appear to be clustered into three episodes separated by 5-10 million year gaps when no seafloor was being generated in the region. Figure 4 illustrates this point and tabulates some of the known major tectonic events that can be

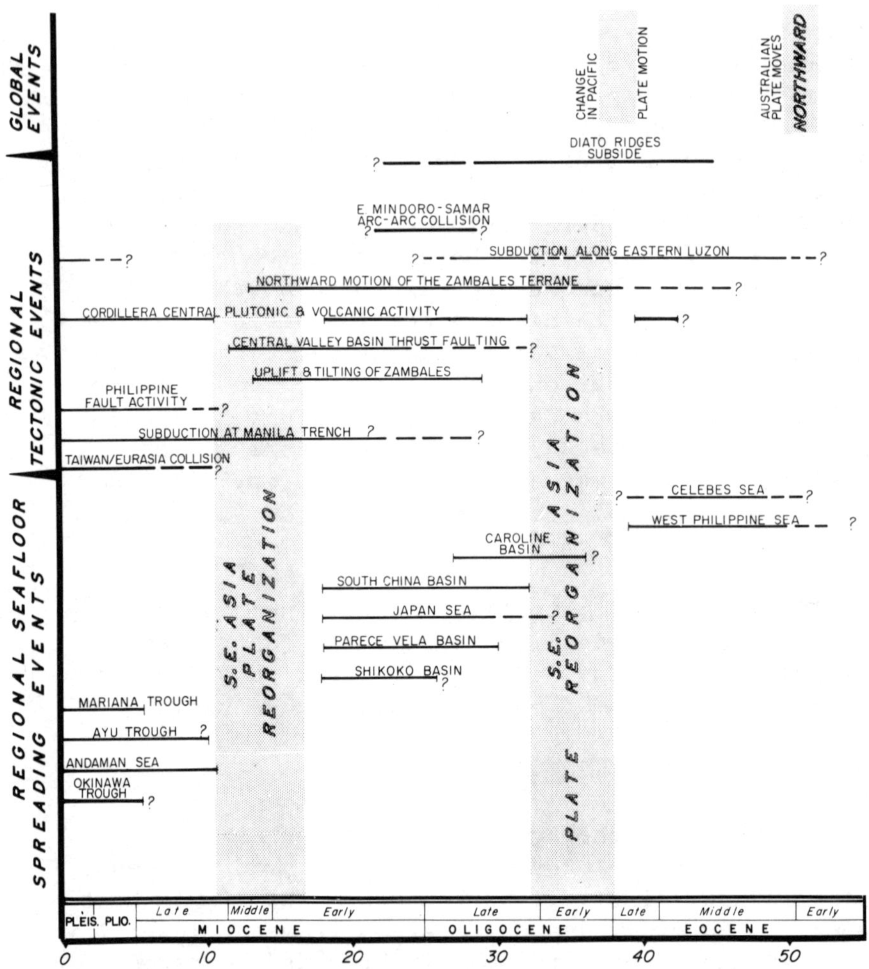

FIGURE 4. Selected tectonic events and known seafloor spreading episodes in Southeast Asian seas. Modified from Hayes and Lewis (in press). Periods of seafloor generation in Southeast Asian marginal seas represent times of active relative plate motions and should punctuate periods of collision related tectonic events in the region.

recognized in the area. At present, the uncertainty regarding the exact duration of these spreading episodes and, more particularly, the time of commencement and cessation of major tectonic events as recorded in the terrestrial geological record are not well established. Hence, a major ambiguity results in trying to establish the existence and nature of any direct cause and effect relationship between tectonic events and episodic pulses of marginal basin development.

Hayes (1982) speculated that pulses of spreading in the region should be manifested as recognizable geological and tectonic events in the surrounding island arcs. Unfortunately, the nature of the time lag between such spreading pulses and some clearly identifiable terrestrial tectonic response to them is not known. Initiation and cessation of spreading episodes provides the clearest evidence of important changes in plate motion in the region. Because the plates are of small dimensions, it is possible that major changes in their relative motions could be triggered by rather subtle changes in plate directions of the much larger, major ocean plates. The generation of new oceanic crust within the area must punctuate the approximate periods in geological time where major changes in the interaction at collisional plate boundaries also occurs.

Hayes (1982) argued that periods of seafloor generation in the area probably represent times of maximum interplate crustal shortening and, hence, should correspond with the most intense tectonic and orogenic activity recorded in the island arcs. Alternatively, Uyeda & McCabe (1983) have speculated that the encroachment at collisional plate boundaries of large, buoyant features "imbedded" in the seafloor may be adequate to inhibit the motion of the converging plate. Hence, such collisions would trigger periods of slowing or no plate motions.

On the one hand, the spreading episodes may predict the periods of tectonism through plate collision which should correspond roughly in time. On the other hand, Uyeda & McCabe (1983) would have the occurrence of major collisions as marking the beginning of quiescent periods in the spreading history of the region. It is not likely that this controversy will be resolved by further examination of the marine record. However, structural and stratigraphic mapping on land, accompanied by better dating of the rock units, provide the best promise for establishing the precise timing of major tectonic events and testing of the above hypotheses.

Our studies related to the SEATAR project have demonstrated that there is considerable along-strike variability in the morpho-tectonic units and structural style manifested along presumed two-dimensional structural lineaments, (e.g., Hayes and Lewis, in press, Lewis and Hayes, in press). A final word of caution is offered to those who might attempt to extrapolate the structural and geophysical data shown in Figure 2 over any significant distances away from the position of the transect zone.

Because the scale of structural and tectonic variations in Southeast Asian seas appears to be small compared with the distribution of our data, we have not yet been successful in establishing any useful paradigm to relate the tectonic history of the region with the development and distribution of metalliferous ore or hydrocarbon resources throughout this region. This will remain one of the important goals of future investigation of this area for some time to come.

ACKNOWLEDGMENTS

This research was supported under NSF grants OCE79-19069 and OCE80-18693. I acknowledge the important contributions of S. Lewis, C. Mrozowski, R. Jarrard, and B. Taylor in the initial compilation of the structure section shown in Figure 2. Furthermore, many of the ideas expressed here represent the ideas expressed in joint publications with the above named scientists and with others. The SEATAR program has enjoyed close cooperation of the CCOP organization, and the important contribution of scientists of the

Philippine Bureau of Mines has been especially valuable for this particular transect zone study. I thank Drs. J. K. Weissel and S. Lewis for their critical review and suggestions for improving this manuscript.

REFERENCES

Anderson, R. N. (1975). Heat-flow in the Mariana Basin. Jour. Geophysical Res. 80, 4043-4048.

Bibee, L. D., Shor, G. G., Jr., and Lu, R. S. (1980). Inter-arc spreading in the Mariana Trough. Marine Geology 35, p. 183-197.

Cooper, A. K., Marlow, M. S., and Scholl, D. W., (1976). Mesozoic magnetic lineations in the Bering Sea marginal basin. Jour. Geophysical Res. 81, 1916-1934.

Hamburger, M. W., Cardwell, R. K., and Isacks, B. (1983). Seismotectonics of the Northern Philippine Island Arc. In: The Tectonic and Geologic Evolution of Southeast Asian Seas and Islands, D. E. Hayes (ed). Geophysical Monograph #27, American Geophysical Union, Washington, D.C., 1-22.

Hayes, D. E., (1983a). Margins of the Southwest Sub-basin of the South China Sea--A Frontier Exploration Target? (Abstract submitted at the Second Workshop on the Geology and Hydrocarbon Potential of the South China Sea and Possibilities of Joint Development, East-West Center, Honolulu, Hawaii. [Manuscript in press, Energy, The International Journal.]

Hayes, D. E. (1983b). Global Studies of Age-Depth Relationships (abs.). EOS 64, 760.

Hayes, D. E. (1982). Magnetic lineations in the marginal seas of Southeast Asia and their tectonic implications. In: Geology and Tectonics of the Luzon-Marianas Region, Proceedings of the CCOP-IOC SEATAR Workshop, Manila, Philippines, 21-28 November, 1981, G. R. Balce & Z. S. Zanoria (eds). Philippine SEATAR Committee Special Publication No. 1, 21-30.

Hayes, D. E. and Caress, D. (in preparation). Depth-Age Relationships in Marginal Sea Basins.

Hayes, D. E. and Lewis, S. D. (in press). A Geophysical Study of the Manila Trench, Luzon, Philippines, Part I: Crustal Structure, Gravity and Regional Tectonic Evolution. Jour. Geophys. Res.

Hayes, D. E. and Ringis, J. (1973). Seafloor spreading in the Tasman Sea. Nature 243, 454-458.

Hayes, D. E., Houtz, R. E., Jarrard, R. D., Mrozowski, C. L., and Watanabe, T. (1978). Crustal Structure. In: Geophysical atlas of east and Southeast Asian seas, compiled under the direction of D. E. Hayes. Geol. Soc. of America Map & Chart Series MC-25.

Hegarty, K. A. and Weissel, J. K., (in preparation). The Caroline Plate - Basins and Boundaries.

Hegarty, K. A., Weissel, J. K., and Hayes, D. E. (1983). Convergence at the Caroline-Pacific Plate Boundary: Collision and Subduction. In: The Tectonic and Geologic Evolution of Southeast Asian Seas and Islands, Part II, D. E. Hayes (ed). Geophysical Monograph #27, Amer. Geophysical Union, Washington, D. C. 326-348.

Hilde, T. W. C. and Lee, C-S. (in press). Origin and Evolution of the West Philippine Basin: A New Interpretation. Tectonophysics.

Hobart, M. A., Anderson, R. N., and Uyeda, S. (1979). Heat transfer in the Mariana Trough (abs.). EOS 60, 383.

Houtz, R. E. and Ewing, J. (1976). Upper Crustal Structure as a function of Plate Age. Jour. Geophysical Res. 81, 2490.

Hussong, D. M. (1981). Underway geophysics--Leg 60 and related surveys. In: Init. Reports DSDP V.60, Hussong, D. M., Uyeda, S. et al. Washington D. C., (U. S. Govt. Printing Office) 71-74.

Hussong, D. M. and Fryer, P. (1981). Structure and Tectonics of the Mariana Arc and Fore-arc: Drillsite Selection Surveys. In: Init. Reports DSDP V. 60, Hussong, D. M., Uyeda, S. et al. Washington, D.C., (U. S. Govt Printing Office) 33-44.

Hussong, D. M., and Sinton, J. B. (1983). Seismicity Associated With Back Arc Crustal Spreading in the Central Mariana Trough. In: The Tectonic and Geologic Evolution of Southeast Asian Seas and

Islands: Part 2, D. E. Hayes (ed). Geophysical Monograph #27, Amer. Geophysical Union, Wash., D. C. 217-235.

Hussong, D. M. and Uyeda, S. (1981). Tectonic Processes and the History of the Mariana Arc: A Synthesis of the Results of Deep Sea Drilling Project Leg 60. In: Init. Reports DSDP V. 60, Hussong, D. M., Uyeda, S., et al. Wash., D.C., (U. S. Govt Printing Office) 909-929.

Hussong, D. M, Uyeda, S., et al. (1981). Initial Reports DSDP V.60. Wash., D.C., (U. S. Govt. Printing Office).

Karig, D. E. (1974). Evolution of arc systems in the western Pacific. Ann. Rev. Earth Planet. Sci. 2, 51-75.

LaTraille, S. L. and Hussong, D. M. (1980). Crustal Structure Across the Mariana Island Arc. In: The Tectonic and Geologic Evolution of Southeast Asian Seas and Islands, D. E. Hayes (ed). Geophysical Monograph #23, Americal Geophysical Union, Wash., D. C. 209-222.

Langseth, M. G. and Mrozowski, C. L. (1980). Geophysical Surveys for Leg 59 Sites, Deep Sea Drilling Project. In: Initial Reports DSDP V. 59, (1981), Kroenke, L., Scott R. et al. Wash., D.C. (U. S. Govt. Printing Office) 487-502.

Lewis, S. D. and Hayes, D. E. (1980). The structure and evolution of the Central Basin Fault, West Philippine Basin. In: The Tectonic and Geologic Evolution of Southeast Asian Seas and Islands, D. E. Hayes (ed). Geophysical Monograph #23, American Geophysical Union, Wash., D. C., 77-88.

Lewis, S. D. and Hayes, D. E. (1983). The Tectonics of Northward Propagating Subduction Along Eastern Luzon, Philippine Islands. In: The Tectonic and Geologic Evolution of Southeast Asian Seas and Islands: Part II, D. E. Hayes (ed). Geophysical Monograph #27, American Geophysical Union, Wash., D. C. 57-78.

Lewis, S. D. and Hayes, D. E. (in press). A Geophysical Study of the Manila Trench, Luzon, Philippines, Part II: Forearc Basin Structural and

Stratigraphic Evolution, Jour. Geophysical Res.
Lewis, S. D., Mrozowski, C. L., and Hayes, D. E. (1982). The origin of the West Philippine Basin by inter-arc spreading. In: Geology and Tectonics of the Luzon-Marianas Region, Proceedings of the CCOP-IOC SEATAR Workshop, Manila, Philippines, 21-28 November, 1981, G. R. Balce and Z. S. Zanoria (eds). Philippine SEATAR Committee, Special Publication No. 1, 31-52.
Louden, K. E. (1976). Magnetic anomalies in the West Philippine Basin. In: The Geophysics of the Pacific Ocean Basin and its Margins, G. H. Sutton et al. (eds). Geophysical Monograph #19, American Geophysical Union, Wash., D.C. 253-267.
Ludwig, W. J. (1970). The Manila Trench and West Luzon Trough - III. Seismic-refraction measurements. Deep Sea Res. 17, 553-571.
Ludwig, W. J., Kumar, N., and Houtz, R. E. (1979). Profiler-sonobuoy measurements in the South China Sea Basin,. Jour. Geophysical Res. 84, 3505-3518.
Meijer, A.,Anthony, E., Reagan, M. (1981). Petrology of volcanic rocks from the fore-arc sites. In: Initial Reports DSDP V. 60, D. M. Hussong, S. Uyeda, et al., (ed). Wash., D.C. (U. S. Government Printing Office) 709-730.
Mrozowski, C. L. and Hayes, D. E. (1979). The evolution of the Parece Vela Basin, eastern Philippine Sea. Earth & Planet. Sci. Letts. 46, 49-67.
Mrozowski, C. L. and Hayes, D. E. (1980). A seismic reflection study of block faulting in the Mariana Forearc Basin. In: The Tectonic and Geologic Evolution of Southeast Asian Seas and Islands, D. E. Hayes (ed). Geophysical Monograph #23, Amer. Geophysical Union, Washington, D. C., 223-234.
Mrozowski, C. L., Lewis, S. D., and Hayes, D. E. (1982). Complexities in the tectonic evolution of the West Philippine Basin. Tectonophysics 82, 1-24.
Parsons, B. and Sclater, J. G. (1977). An analysis of the variation of ocean floor bathymetry and heat flow with age. Jour. Geophysical Res. 83, 803-827.
Scott, R. B., Kroenke, L., Zakariandze, G., and Sharaskin, A. (1980). Evolution of the South

Philippine Sea: Deep Sea Drilling Project Leg 59 results. Initial Reports DSDP V. 59, L. Kroenke and R. Scott et al. (eds). Wash.,(U. S. Government Printing Office) 803-816.

Shih, T-C. (1980). Marine Magnetic Anomalies from the Western Philippine Sea: Implications for the Evolution of Marginal Basins. In: The Tectonic and Geologic Evolution of Southeast Asian Seas and Islands, D. E. Hayes (ed). Geophysical Monograph #23, Amer. Geophysical Union, Washington, 49-76.

Taylor, B. and Hayes, D. E. (1980). The Tectonic Evolution of the South China Basin. In: The Tectonic and Geologic Evolution of Southeast Asian Seas and Islands, D. E. Hayes (ed). Geophysical Monograph #23, American Geophysical Union, Wash.,D.C. 89-104.

Taylor, B. and Hayes, D. E. (1983). Origin and History of the South China Basin. In: The Tectonic and Geological Evolution of Southeast Asian Seas and Islands: Part II, D. E. Hayes (ed). Geophysical Monograph #27, Amer. Geophysical Union, Washington, D. C., 23-56.

Taylor, B. and Karner, G. D. (in press). On the Evolution of Marginal Basins. Rev. Geophys. Space Phys.

Uyeda, S. and Ben-Avraham, Z. (1972). Origin and development of the Philippine Sea. Nature Phys. Sci. 240, 176-178.

Uyeda, S. and Horai, K. (1981). Heat Flow Measurements on Deep Sea Drilling Project Leg 60. In: Initial Reports DSDP V. 60, D. M. Hussong, S. Uyeda, et al. (eds). Washington (U. S. Government Printing Office) 789-802.

Uyeda, S. and McCabe, S. (1982). A possible mechanism of episodic spreading of the Philippine Sea. In: Geology and Tectonics of the Luzon-Marianas Region, Proceedings of the CCOP-IOC SEATAR Workshop, Manila, Philippines, 21-28 November 1981, G. R. Balce & A. S. Zanoria (eds). Philippine SEATAR Committee, Special Publication No. 1, 53-66.

Watts, A. B., Weissel, J. K. & Larson, R. L.(1977). Sea floor spreading in marginal basins of the

western Pacific. Tectonophysics 37, 167-181.
Weissel, J. K. (1981). Magnetic lineations in marginal basins of the west Pacific. Phil. Trans. R. Soc. Lond. A. 300, 223-247.
Weissel, J. K. and Anderson, R. N. (1978). Is there a Caroline Plate? Earth Planet. Sci. Letts. 41, 143-158.

Lamont-Doherty Geological Observatory Contribution Number: 3626

OBSERVED STRAIN AND THE STRESS GRADIENT ACROSS SOME FOREARC AREAS OF MODERN CONVERGENT MARGINS

ROLAND VON HUENE

U.S. Geological Survey
Menlo Park, California 94025

ABSTRACT

Compressional stress from plate convergence transmitted across a modern subduction zone decreases towards land and becomes tectonically insignificant in many forearc basins. Intense Pliocene to Holocene compressional tectonism in the trench slope falls off landward and much structure in seismic records across several forearc basins is non-compressional to depths of 5-10 km. The rapid landward decrease in compressional deformation can be accomplished by a reduction of mechanical friction between the upper and lower plates in a subduction zone. There are several ways to reduce friction but one way that is potentially present in most subduction zones is the decoupling due to overpresssured pore fluids. Such decoupling may continue to depths greater than 100 km in the subduction zone because of both interstitial water in the upper subducted crust and water made available through mineral dehydration reactions in response to increasing temperatures. When water is trapped at depth, pore pressures can be elevated to lithostatic levels. This reduces friction along subduction zones to the low levels required by the extensional structure along some arcs, forearcs, and opening backarc basins.

Proceedings of the 27th International
Geological Congress, Volume 23, pp. 155–188
ORIGIN AND HISTORY OF MARGINAL AND INLAND SEAS

INTRODUCTION

A junction between oceanic and continental crust marked by an arc-trench system is a convergent margin, so-called because it is formed by the convergence of two lithospheric plates. Research during the past decade has independently confirmed the implications from plate tectonic theory that the crust generated at mid-ocean ridges is disposed of by subduction at convergent margins. Modern seismic reflection records and unbroken oceanic magnetic anomalies beneath the edge of the continent that reveal the continuation of ocean crust for many tens of kilometers have confirmed subduction of the oceanic plate. Direct evidence that the Be^{10} isotope in sediment overlying subducting ocean crust has traveled hundreds of km beneath the continent is revealed by relatively high concentrations of Be^{10} in lavas of the Aleutian and Middle American volcanic arcs (Brown et al, 1982). The recognition of far-traveled continental fragments incorporated into the Pacific borders of much of North America requires that considerable oceanic crust has been subducted since the beginning of Mesozoic time.

Although assumed to be a basic geologic process, convergence has several puzzling aspects. Rapid rates of convergence provide potential for considerable compressional stress along subduction zones, but structure in the upper plate indicates that compressional strain is commonly mild or unrecognizable landward of the trench slope. The association off extension in the form of a spreading backarc basin that is related in some fashion to the compressional subduction process, seems paradoxial. It appears that commonly, the dominant stress field in the upper plate changes from compression to non-compressional somewhere in the forearc area.

An arc-trench system with a spreading backarc ridge has 2 plate boundaries (fig. 1), the subduction zone, and the spreading ridge; the intervening arc terrain is an independent plate (Bird, 1978). Compression should result from shear in the subduction zone, whereas compression must be very low or absent in the spreading ridge. If the arc is not accelerating, the horizontal forces perpendicular to the arc will balance. By presuming balanced horizontal forces, Bird (1978) calculated much lower stresses across the Mariana and Tonga subduction zones than reported from previous model studies. He postulated that pore fluid pressures nearly equal to the lithostatic load decreased stresses in the subduction zone.

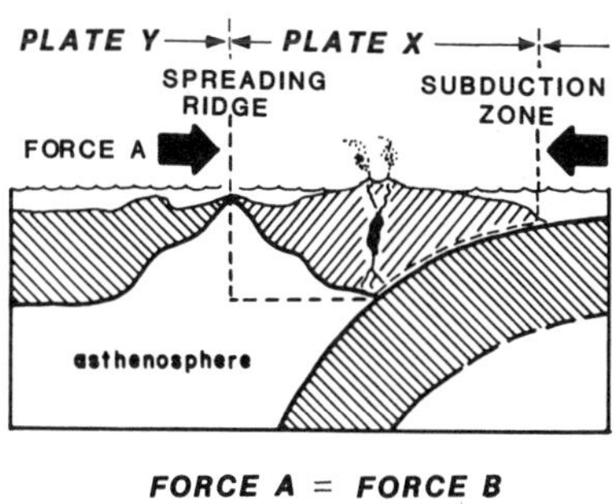

Figure 1. Diagram across an arc-trench system with seafloor spreading in the backarc basin. The subducting and spreading plate boundaries isolate the arc terrane as a separate plate (plate X) across which the forces on each plate boundary must balance (force A = force B) (from Bird, 1978).

The importance of compressional deformation may have been exaggerated in quite a few studies of modern subduction zones. Although decoupling may be mentioned, the extent of decoupling and apparent weakness of material in the accretionary complex is often not discussed as much as the compressional deformation. A lack of resolution in seismic records has permitted a bias toward compressional structure as is evident in some areas where single channel data were superceded with well processed multichannel data. This is not to imply that subduction zones are without widespread compressional tectonism; parts of the Andean margin are examples where compressional zones occur behind the arc despite the intervening extensional zones (Jordan et al, 1983; Suarez et al, 1983). The spectrum of stress gradients across convergent margins encompasses the compressional Andean and more extensional Mariana examples (Uyeda, 1982).

This paper's focus is on the extent of <u>active</u> compressional deformation across several forearc areas as depicted in recently gathered multichannel seismic reflection records. Evidence for compression comes from folding, reverse, and thrust faulting which are measures of strain or deformation. Near vertical and steep faults are refered to as non-compressional. For this discussion, I divide the arc-trench system (fig. 2) into:

1) the trench
2) the lower part of the trench landward slope (lower slope)
3) the upper part of the trench landward slope (upper slope)
4) the shelf edge structure or uplift
5) the forearc basin
6) the arc

Gridded multichannel seismic reflection surveys and drill information in the forearc basins of the Aleutian, Japan, Mariana and Middle America, margins illustrate the distribution of compressional and non-compressional deformation produced within a geologically short period of time (2-4 m.y.). The age of deformation is an important factor since accretionary complexes develop progressively. The youngest and most rapid deformation is near the trench and deformation becomes older and less rapid landward. Eventually structure becomes fixed and

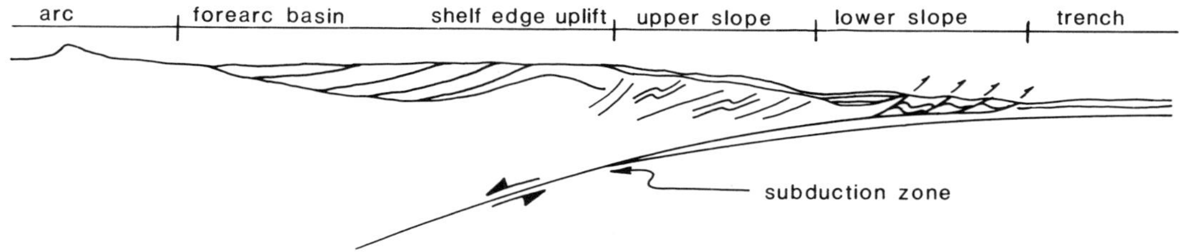

Figure 2. Diagram showing terminology for arc-trench features used in this paper.

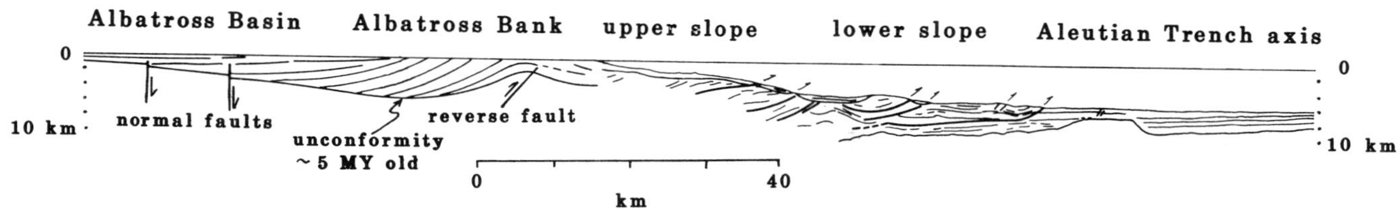

Figure 3. A cross section based on multichannel seismic records from Albatross Basin to the Aleutian Trench converted to depth and without vertical exaggeration (after Fisher and von Huene, 1980; and von Huene et al, 1983). This example is shown in depth rather than time to illustrate a section at true scale across the front of the Aleutian margin.

inactive as a zone is uplifted, becomes part of the upper slope, and the magnetude of horizontal compression decreases to insignificant levels. Thus, the deformation may have a different rate and style between the base and the top of a slope despite the similarity of structure in both areas.

EXAMPLES OF STRUCTURE

Aleutian Trench

The forearc and slope of the eastern Aleutian Trench of the Kodiak and Shumagin shelves have been well surveyed with multichannel seismic reflection, and other geophysical techniques (Fisher and von Huene, 1980; Fisher, 1979; von Huene et al, 1979; Bruns and von Huene, 1977). DSDP drilling was accomplished on Leg 18 (Kulm, von Huene et al, 1973) and conventional sampling and industry drilling have been summarized (Keller et al, 1984 in press). Studies of these data indicate that the present tectonic system began in the latest Miocene, concurrent with an increase in explosive arc volcanism (Sheiddegger, 1980). At this time much of the forearc area was a broad erosional surface which subsided differentially and formed 6 local sediment-filled forearc basins 3 to 9 km deep. The unique shape of each basin most likely reflects reactivation of pre-late Miocene structure.

Albatross Basin off southern Kodiak Island has a steep seaward flank against a structural uplift, Albatross Bank, and a gentle landward flank (fig. 3). Albatross Bank is a folded structure at the edge of the shelf; asymmetric folds are offset by small reverse faults, and the entire bank has been uplifted at least 3,000 m since the late Miocene. Albatross Bank has experienced more compressional deformation than adjoining areas and seaward of the bank, upper-slope deformation is principally uplift. The landward-dipping beds, comprising much of the slope were tilted in pre-Pliocene time, are covered by a sediment blanket only moderately disrupted by faults. The most severe compressional deformation has occurred on the lower slope in the form of thrust faults and tight folds (von Huene et al, 1983) (fig. 3) whereas mildly extensional deformation characterizes the area landward of Albatross bank as indicated by steep normal faults (Fisher and von Huene, 1980). The transition from compression through little deformation to extensional deformation occurs landward of Albatross Bank.

This pattern of compression seaward and extension landward typifies many basins in the region. Tugidak Basin (Fisher, 1979) adjacent to Albatross Basin, has a similar division of normal and reverse faulting (fig. 4). Sanak Basin (Bruns and von Huene, 1977) is cut by vertical and extensional faults (fig. 5) superimposed on a strong pre-existing transverse structural grain at 45, to the trend of the arc. Shumagin Basin, however, exhibits little deformation from horizontal stress (fig. 6). Seaward of the Shumagin and Sanak basins the slightly-deformed upper slope is covered by a relatively uniform sediment sequence. Seismic records reveal intense compressional deformation only across the lower slope (Bruns et al, 1983).

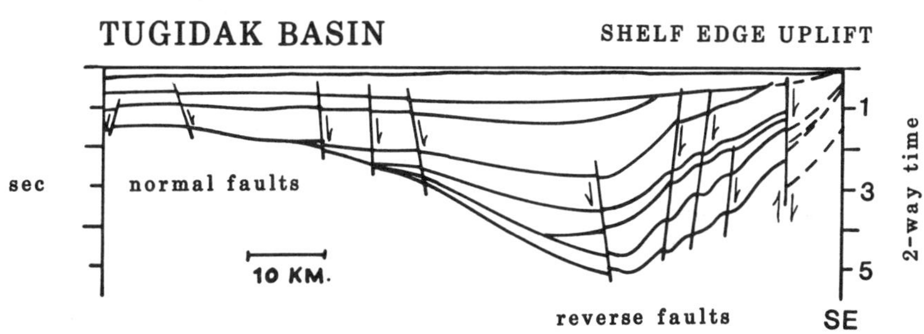

Figure 4. Tugidak Basin (vertical exaggeration 4x) showing small normal faults on the landward flank and small reverse faults on the seaward flank. The basin began to fill with sediment five m.y. ago, concurrent with faulting, but most faults are buried beneath Pleistocene sediment (from Fisher, 1979).

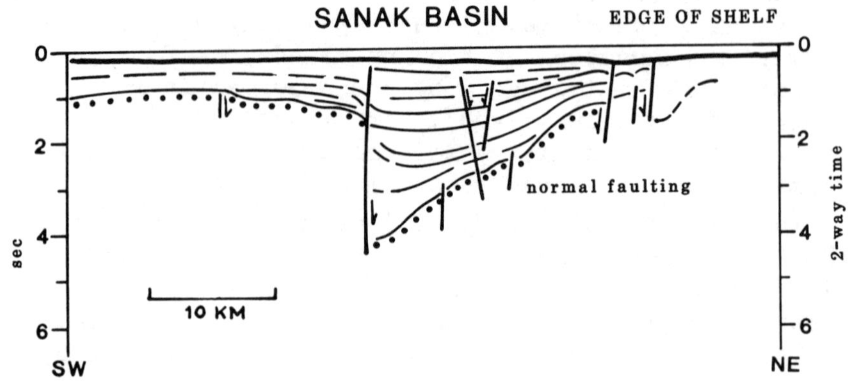

Figure 5. Sanak Basin, (vertical exaggeration 4x) a half graben structure with a long axis oriented transverse to the shelf and sub-parallel to the present direction of relative motion between the North American and Pacific plates. The axial trend of the basin parallels the strike of the Bering shelf indicating reactivation of deformation along an ancient structural direction (from Bruns and von Huene, 1977).

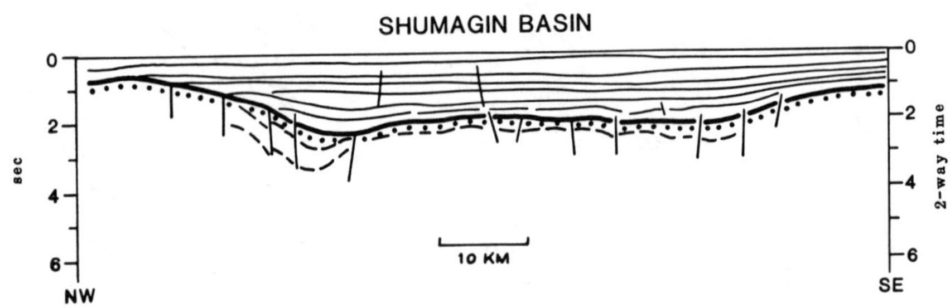

Figure 6. Shumagin Basin, (vertical exaggeration 4x) a little deformed nearly circular basin between Tugidak and Sanak Basins (from Bruns and von Huene, 1977).

A comprehensive suite of data has recently been collected in the Amlia corridor (173,W longitude) of the central Aleutian arc (Scholl et al, 1982; McCarthy and Scholl, in press). The lower slope consists of deformed trench sediment including monoclinal tilted sequences, folds, and extensive as well as short thrust packets. The adjacent forearc basin contains Oligocene through Holocene volcanic debris up to 5 km thick derived from the adjacent arc. Both the forearc sediment and the Eocene and older volcanic rock of the arc are little deformed.

A major difference between the eastern and central Aleutian arcs is that the first is adjacent to a continent whereas the second is intra-oceanic. The central Aleutian arc has a very narrow continental shelf, and the forearc basin is in 4000 to 5000 m of water. Otherwise the distribution of compressively deformed strata are very similar with little or no deformation in the forearc basin but considerable deformation on the lower slope.

Japan Trench

A segment of the Japan trench has been extensively surveyed with single channel techniques (Honza et al 1977), multichannel industry surveys (Ishiwada and Ogawa, 1976), and contract multichannel records for academic purposes (Nasu et al, 1980). This was followed by DSDP drilling on Legs 56-57 (Scientific Party, 1980). A series of low folds were infered to be compressional features from the high vertical exaggeration in single channel data; however, the multichannel data depicts many of these folds as shallow features above normal and occasional high angle reverse faults not reaching the seafloor (fig. 7). In record JNOC-1, a sequence of small (vertical displacement 100 m or less) normal faults is associated with the uplift of a broad knoll and may thus be related to extension in only the sediment cover over the knoll. However, normal faults in records far from the knoll are interspersed with occasional small vertical faults and small reverse faults.

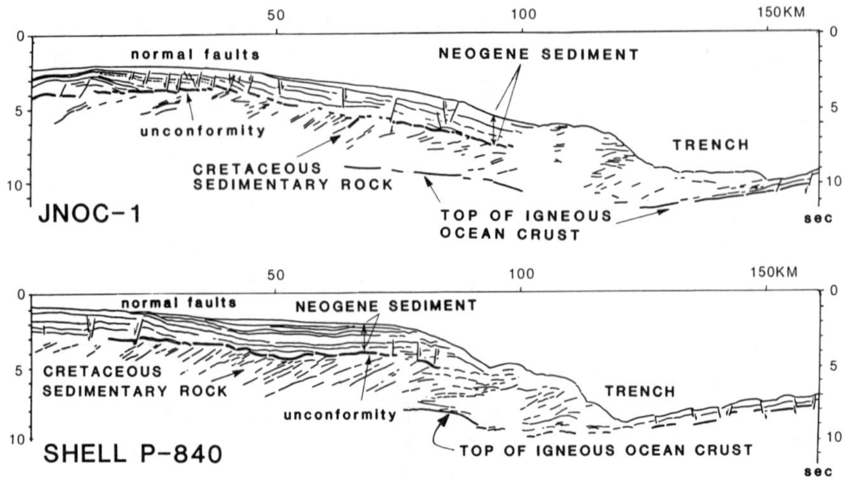

Figure 7. Two sections across the seaward part of the Japan Trench margin (vertical exaggeration 9x) showing many small normal faults in a non-disrupted Neogene sediment sequence. The Neogene sequence is unconformably underlain by slightly metamorphosed Cretaceous rocks. The ocean crust appears deflected upward because of the distortion in a time presentation.

The faults rarely produce scarps at the seafloor and the Pliocene and the Quaternary section is little deformed. This lack of Quaternary faulting is coincident with the change in stress across Honshu Island expressed offshore as a change from subsidence to uplift about 3 m.y. ago (Keller, 1980). Thus the stress field in which the normal faults were formed is no longer active. Normal faulting was coincident with Neogene arc-volcanism and is thus infered to be coincident with subduction (von Huene et al, 1982). The normal and some reverse faults occur beneath the continental terrace and the top of the upper slope. Seismic data from the lower slope, especially that part below the mid-slope terrace, is generally without many coherent reflections and the strata are probably deformed as suggested by DSDP drill cores (Scientific Party, 1980; von Huene et al, 1982).

Along the Japan Trench, much of the Miocene structure of the continental terrace and upper slope is extensional or non-compressional. The zone of stratal disruption and poor seismic resolution is concentrated along the lower slope.

Middle America Trench, Guatemala

A segment of the shelf, slope and trench of the Middle America Trench off Guatemala was extensively surveyed with a multichannel seismic system (Ladd et al, 1982) following an earlier report based on industry data (fig. 8) (Seely et al, 1974, Seely 1979). The area was drilled during 2 DSDP legs (Aubouin et al, 1982a). The shelf is underlain by a Cretaceous through Paleocene base of slope sediment sequence upon which rests a Eocene through Quaternary forearc basin sequence. The forearc sequence is little deformed, but is cut locally by steep normal and reverse faults, many of which do not extend to the seafloor (Ladd, in press; von Huene, in press).

The seaward edge of the forearc basin has been uplifted but within the basin filling sediment, faulting is minor. The upper and mid-slope regions are covered by slope sediment which is poorly stratified. The underlying basement is very rough and probably faulted (zone of poorly defined faulting, fig. 8). Locally the foot of the slope maybe marked by collapse (von Huene, in press), and Aubouin et al (1984, in press) have emphasized the extensional character of the landward slope of the trench.

Compressional deformation on the lower slope off Guatemala is not clearly shown away from the DSDP transect. Along the transect, the lower slope collapse obscures much of the older structure. Accretion may mark the base of the slope in the record published by Seely et al (1974). However, from the upper slope landward, the Eocene to Quaternary forearc stratified section is little deformed and vertical offsets along faults are generally less than 100 m.

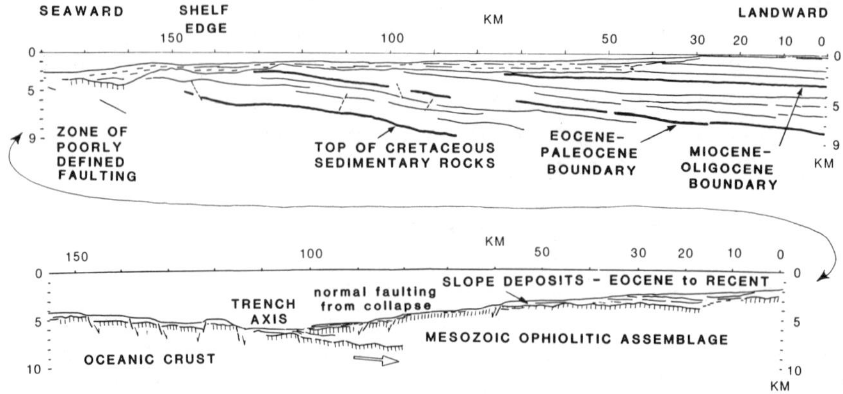

Figure 8. Cross section of the Middle America Trench margin off Guatemala. Vertical axis is in depth and vertical exageration is 2:1. The upper half of the section shows the forearc basin underlain by a prior base of slope sequence of Cretaceous and Paleocene age (Seely, 1979; von Huene, in preparation). The sediment sequence is little deformed, but ends seaward at a poorly known zone of deformation beneath the upper slope covered by slope deposits. The slope is underlain by an ophiolitic sequence (Aubouin et al, 1982) and in the area of the IPOD transect, the lowermost part has failed by collapse. No accretionary complex was found along this transect which was drilled on DSDP Legs 67 and 84.

Mariana Trench

Geophysical surveys across the Mariana forearc near 18,N have outlined thick strata that depen toward a basin axis near the volcanic arc in a western province, and a thin sequence of sediment in an eastern province. Both provinces are broken by many normal faults (fig. 9) (Mrozowski and Hayes, 1980; Hussong and Fryer, 1982). The Eocene to Quaternary sediment overlies an igneous basement of arc related rock and was drilled on DSDP Leg 60 (Hussong, Uyeda et al, 1982). Structures suggesting compression are rare or absent across the terrace trenchward of the volcanic arc and mostly tensional faults were found in seismic records of the transect surveyed. The faults generally strike parallel to the arc and have vertical offsets from 10 to 200 or 300 m with downward displacement toward the trench. No structural uplift marks the edge of the shelf but large seamounts thought to be serpentinite diapers are present in the forearc area

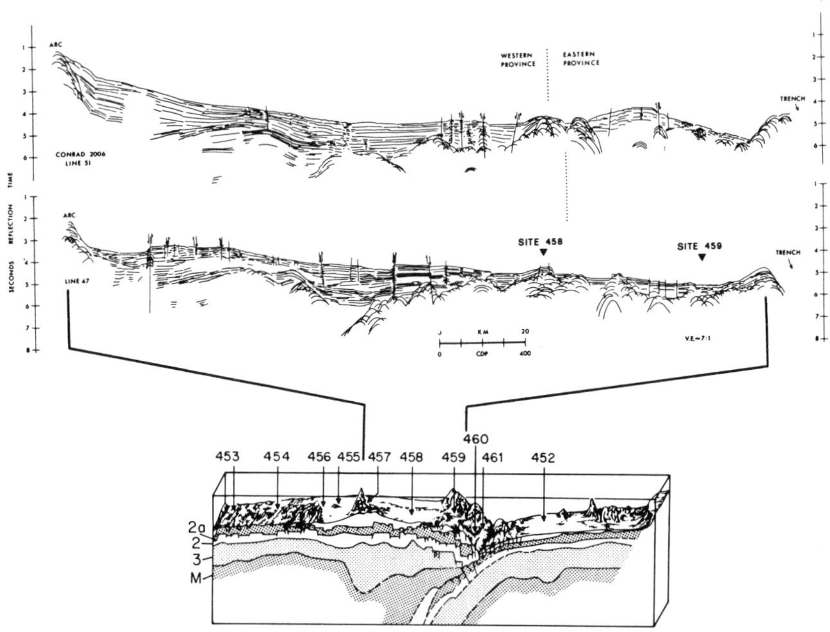

Figure 9. The Mariana forearc area showing many small normal faults in a sediment section of Eocene to Holocene age (from Mrozowski and Hayes, 1980).

(Hussong and Fryer, 1983). The seaward limit of normal faulting extends at least to the upper part of the landward slope of the trench.

Other Margins

Along some other less-extensively studied margins the landward extent of active compressional deformation does not extend far beyond the seaward edge of the forearc basin. Near Peru, the unfaulted basinal or slope sediment inferred to be of Neogene age (Hussong and Wipperman, 1981, Kulm et al, 1981) is located about 50 km from the active accretion near the trench. Off Nias Island in the Sunda Arc, most of the Pliocene to Quaternary compressional deformation has occurred on the lower slope (Karig et al, 1980). Several seismic records show that the forearc area of the Tongan Arc is normally faulted to the edge of the shelf (Lehner et al, 1983).

The examples of deformational style across the convergent margins cited are not an exhaustive or systematic listing. They do, however, demonstrate that along several margins where subduction is unaffected by collision, active compressional deformation is typically intense on the lower slope and only occasionally extends landward of the top of the trench slope. When compressional deformation occurs at the top of the slope it is associated with a structural uplift marking the juncture of the slope and shelf. In these forearc basins, strata are typically cut by near-vertical faults of non-compressional character. Local reverse faults are observed in association with transverse features and are seen in rocks deformed in a prior period of deformation.

RAPID INITIAL DEFORMATION ACROSS THE FOREARC

A distinct difference between the intense deformation across the lower slope and less active deformation across the upper slope became evident as the resolution of structure in seismic records improved and as more margins were drilled and sampled by DSDP.

The first to semiquantitatively estimate the initial high rate of deformation were Moore et al (1981). A well-processed seismic record across the Middle America Trench off Mexico (Shipley et al, 1980) shows rotation of reflectors at the base of the slope to the 25,-30, landward dipping pattern characteristic of many convergent margins. However, rather than the gradual steepening depicted in models, the bulk of the rotation of reflections occurred in the first few kilometers at the base of the trench slope, and was accomplished in less than 1 m.y. Moore and his colleagues (1981) plotted age versus the predominent dip of bedding in cores from the DSDP Leg 66 transect of drill holes. They recognized a dramatic 90% decrease in the rate of tilting from the base of the slope 10 km landward where stress from the subduction zone is transmitted through only 1 to 1.5 km of sediment. The qualitative and quantitative evidence indicate a concentration of deformation in accreted sediment at the front of the margin and a rapid rate of decreasing deformation away from the zone of underthrusting.

Moore and Biju-Duval used a similar approach in interpreting the results of DSDP Leg 78A (1984, in press). From seismic and drill information across the Antilles Arc, these authors made a semiquantative estimate of the distribution of deformation across the first 45 km of the landward slope of that margin. About 75% of the initial thrusting and tilting in the late Miocene to Quaternary accreted sediment section occurs within the seaward half (22 km) of the deformation zone. The rate of convergence here is only about 2 cm/yr as contrast with the 10 cm/yr off southern Mexico.

An estimate of the rate of initial deformation was made across the Nankai Trough where exceptional seismic records (Aoki et al, 1983) and a transect of DSDP drill holes (Karig and Kagami, 1983) provide quantitative constraints. Karig (1984, in press) has used two semiquantitative techniques, one a palenspastic restoration to a balanced crossection, the other a mass balance of input volume to residual volume of sediment to study deformation. Both analyses indicate that about 50% of the deformation is accomplished in the seaward 8 km and about 70% in the seaward 15 km of the accreted sediment. The remaining deformation appears to be distributed across the lower slope of the trench. Here the rate of convergence is about 2 cm/yr or less.

These semiquantitative estimates and the previous observations of structure indicate a zone of intense compressional deformation adjacent to the trench axis and less active compressional deformation up the slope toward the edge of the shelf.

INDICATIONS OF STRESS ACROSS THE ALEUTIAN ARC

The observation of non-compressional structure in Aleutian forearc basin sediment sequences is in apparent conflict with recent proposed stress trajectories along the Aleutian and other convergent areas where an axis of maximum horizontal compression is across the arc (Nakamura et al, 1980, Nakamura and Uyeda, 1980). It does not seem possible that such a horizontal compressional stress gradient could decrease to zero between the trench axis and the seaward edge of the forearc basin, remain at zero across the deep forearc basin to then be reactivated again at the volcanic arc.

A closer examination of the arguments indicating compressional stress in the volcanic arc shows first, that the examples are more complex than first thought and that zones of collision were included in the analysis (c.f. Nakamura et al, 1980; and Nakamura and Uyeda 1980). Collision, especially where thickened crust is subducted, can promote strong coupling across a subduction zone because such crust when subducted has greater than normal crustal buoyancy. The increased coupling of thick continental crust extends the horizontal stress gradient. Because this symposium is concerned with the marginal basins behind a subduction zone, only non-collisional areas are considered here; thus when the collision zones of the Aegean and central European areas (Nakamura and Uyeda, 1980) are excluded, attention turns to the non-collisional part of the

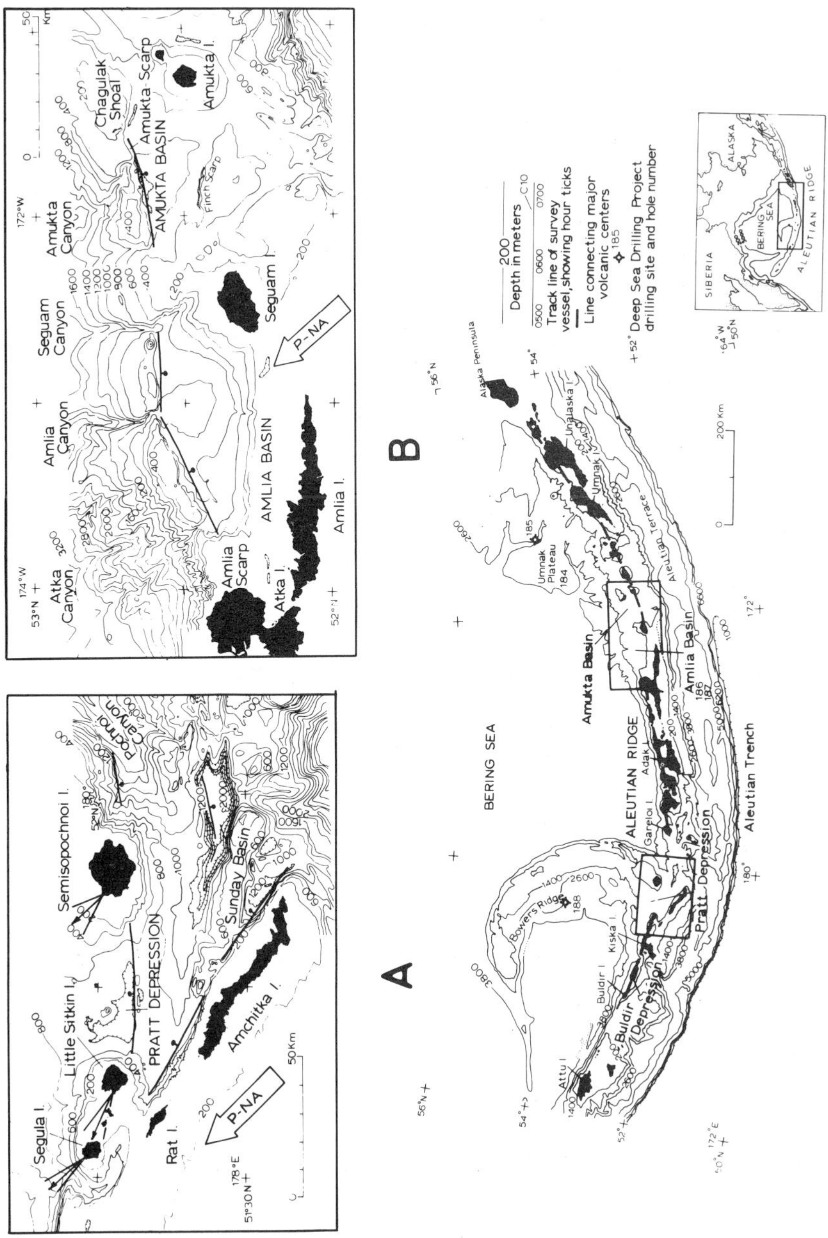

A
Segula I.
Little Sitkin I.
Semisopochnoi I.
PRATT DEPRESSION
Pochnoi Canyon
Sunday Basin
Rat I.
Amchitka I.
P-NA
50Km
178°E
51°30N
B
Atka Canyon
Amlia Canyon
Seguam Canyon
Amukta Canyon
Chagulak Shoal
Amukta Scarp
AMUKTA BASIN
Finch Scarp
Amukta I.
Amlia Scarp
Atka I.
AMLIA BASIN
Amlia I.
Seguam I.
P-NA
174°W
172°W
53°N
52°N
BERING SEA
ALEUTIAN RIDGE
Bowers Ridge
Attu I
Buldir I
Kiska I
Gareloi I
Adak I
Buldir Depression
Pratt Depression
Amukta Basin
Amlia Basin
Aleutian Terrace
Aleutian Trench
Umnak Plateau
Umnak I
Unalaska I
Alaska Peninsula
200 Km
Depth in meters
Track line of survey vessel, showing hour ticks
Line connecting major volcanic centers
Deep Sea Drilling Project drilling site and hole number
SIBERIA
ALASKA
BERING SEA
ALEUTIAN RIDGE

Aleutian Arc along the Alaska Peninsula and the central Aleutian segment (fig. 10). The example of northern Japan is set aside because of the newly proposed initiation of a convergent plate boundary along the Sea of Japan coast (Nakamura, 1983).

In the Aleutian arc, Nakamura et al (1980) have interpreted the regional stress field from the pattern of dikes and the orientation of subsidery cones on composite volcanos. Because of the alignment of the compressional axis with the direction of convergence between the Pacific and North American plates they assume that the maximum compressional axis coincides with the direction of maximum principal stress. This appears to be at variance with the geologic structure of the Aleutian Arc. For example, no compressional structure as a result of Pliocene to Quaternary convergence is known along the volcanic arc, and instead, some of the best studied areas indicate major extensional features. A recent detailed marine geophysical survey of the Amlia corridor (Scholl et al,

---

Figure 10. Two geophysically surveyed areas on the Aleutian Ridge showing normal faults and basins with graben or half-graben structure. In map A, the direction of maximum compressional stress determined from volcanoes on Segula, Little Sitkin and Semisopochnoi Islands are plotted with limits of certainty (Nakamura et al, 1980). The direction of relative motion between the Pacific and North American plates (P-NA) is shown by a bold arrow. Normal faults crossed by at least one seismic line and projected along the bathymetric scarp are shown by heavy line, pendulum symbol on down-thrown side (After Scholl et al, 1975).

1982) includes two arc summit basins which contain Neogene and Quaternary sediment (fig. 10). The basins are elongate sub-parallel to the arc and the sediment section is cut by down-to-basin normal faults.

Some of the normal faults in the area around Semisopochnoi Island (fig. 10a) parallel the maximum compressional stress direction as the analysis of Nakamura et al (1980) would predict; others deviate significantly from the direction and probably follow a previous structural trend. Normal faults in the Amlia corridor are transverse to the maximum compressional stress direction (fig. 10b). These normal faults from 20 km to 65 km long, are interpreted from sharp fault scarps, truncated beds in seismic records, and they are drawn parallel to the orientation of well defined bathymetric scarps (Scholl et al, 1975; Scholl et al, 1982), Basins bounded by normal faults dominate the structure of the arc and the Quaternary basin fill shows no significant compressional structure. Clearly, in the best surveyed areas, the arc is not failing by compression but by extension; thus the axis of principal stress must be tensile and the axis of compressional stress may be vertical in the upper levels of the crust.

Although it is difficult to differentiate much of the Quaternary structure from earlier structure, detailed maps of the Aleutian islands show no dominantly compressional deformation but typically show conjugate fault patterns at angles to the trend of the arc. Exposed early Miocene plutons are roughly equidimensional rather than elongate as expected of intrusions into horizontally compressed rocks (Marsh, 1980; McLean et al, 1980; Vallier, 1983).

On the Alaska Peninsula, recent maps show high-angle reverse as well as non-compression faults which form a structural grain paralleling the arc. However, faults transverse to the arc are numerous and both Veniaminof and Aniakchak volcanos are along the projection of major transverse faults (Detterman et al, 1979, and 1983). These volcanos were intruded into a terrane with a previously established structural grain that does not necessarily correspond to the present stress field. They could have vented along preexisting zones of weakness which control the alignment of volcanic vents and dikes.

Observations of structure along the Aleutian Arc indicates that the determination of stress direction by Nakamura et al (1980), based on the pattern of dikes, flank, and post-caldera cones on composite volcanos, is valid, however, in the upper crust the direction of maximum principal stress
is tensional and that the compressional axis is either the intermediate or the vertical axis.

DECOUPLING ACROSS SUBDUCTION ZONES

The relatively narrow zone of compressional deformation limited to the front of many subduction zones, as compared to the great width of compressional deformation in collision zones on land (e.g. the Himalayan Mountains), indicate that at non-collisional convergent margins the horizontal stress gradient falls off rapidly, and less horizontal compression is transmitted from lower to the upper plates despite high rates of convergence. Perhaps the relatively small amount of transmitted horizontal stress can be attributed to the uniqueness of a "Benioff" type of subduction zone (Bally, 1980).

Using a limited number of earthquakes in the upper plate of subduction zones, Uyeda and Kanamori (1979) proposed that extensional earthquakes are common in backarc basins with spreading ridges (Mariana or M-type) and non-extensional earthquakes are common in non-spreading in backarc basins (Chilean or C-type). They attributed this difference to variable mechanical coupling between the upper and lower plates. Nakamura and Uyeda (1980), however, presented some conflicting data and considered a simple C-type and M-type division as based on incomplete data because additional backarc areas without good focal mechanisms did not fall into a simple two-part classification. On the basis of more complete data they concluded that backarc extension is not associated only with a spreading ridge but that extension affects non-spreading backarc areas as well.

The extent of decoupling along subduction zones

has aroused the interest of many authors. Uyeda (1982) considered several causes of decoupling on a global scale. These include the configuration of the Wadati-Benioff zone, the age, and thus the relative negative buoyancy, of the subducting crust, and the absolute motion of the upper plate, either toward or away from the trench. At a local scale he considered upper mantle wedge convection (McKenzie, 1969; Sleep and Toksoz, 1973), an excess of mass brought in by mantle flow (Bostrom, 1971; Ida, 1978), and temperature anomalies. All of these local mechanisms probably contribute to the complex balance of forces in an arc-trench system at one time or another. The importance of local effects is demonstrated by the variation of structure along single arcs where the global effect of Wadati-Benioff zone configuration, age of the crust, and the rate and direction of plate motion are everywhere the same. Even where global affects are constant, the local structural diversity can span the spectrum from accretion to subduction erosion as along the forearc area of the middle America Trench (Watkins, Moore et al, 1981; Shipley et al, 1982; Aubouin, et al, 1983) and the Aleutian Arc (Scholl et al, 1983; von Huene, 1984 in press). Therefore, the differences in the degree of coupling caused by local effects can equal those of global effects.

Locally the strength or weakness of the coupling across the front of a subduction zone is associated with differences in the thickness and strength of sediment entering a subduction zone and the topographic roughness of the subducting plate. The shear that can be transmitted across the sediment or crushed rock in a zone of thrusting cannot exceed the strength of the interface material. The strength of rocks and sediment of the subduction zone is highly dependent on pore-fluid pressure. Elevated fluid pressure can decrease shear strength more than the differences in material (Hubbert and Rubey 1959). If pore fluid in a subduction zone were elevated from hydrostatic to lithostatic pressures (a pressure equal to the mass of the overlying materials), the upper plate would nearly float upon the lower plate, and the stress transmitted across the subduction zone or the coupling would be weak. Direct measurements of near-lithostatic pressure were made at the front of the Barbados and Guatemalan subduction zones (Moore, Biju-Duval et al 1983, Aubouin, von Huene et al 1982).

Elevated pore-fluid pressure in forearc basins is indicated by pressure measurements in industry drill holes, diapiric structure, and from model studies (Davies et al, 1983; von Huene, 1984 in press). Along the Japan Trench locally elevated pressure in slope sediment has probably developed in fracture porosity (Carson et al, 1981). Where they have been tested, the frontal areas of subduction zones have high pore water pressures.

The amount of interstitial water is very large in accreted or subducted sediment. The porosities measured in trench sediment drilled by the Deep Sea Drilling Project indicate that in the upper kilometer of trench sediment, half is water. Water is also stored in the pores or fractures of sediment and igneous rock in the subducting ocean crust. Thus, the availability of water is probably sufficient to cause mechanical isolation across various complex interfaces if it becomes overpressured.

Pore-fluid pressure levels are a function of two factors: 1) the rate at which pressure is applied, derived from the rate at which horizontal and vertical stresses are applied to the materials in a subduction zone, and, 2) the ability of fluids to drain from the sediment or the rate at which fluid pressure is released. Horizontal pressure results from the convergence of two plates, and vertical pressure occurs due to loading as the thickness of accreted rocks and sediment increases. If the convergence of two plates is very slow, the contained fluids have more time to drain from the material and to release pressure continuously, thus keeping pressures low. An increase in the rate of convergence without a commensurate increase in the rate of drainage results in an elevation of pressure. Rising pressure with increasing convergence rate will weaken the material until the stress exceeds the strength of the material. Once the rock begins to fail, the style of deformation will be affected by differences in pore fluid pressure (effective rock strength).

Accreted sediment becomes progressively dewatered with increasing deformation, age, and burial. Observations of greatly dewatered sediment from a DSDP drill hole at the base of the Aleutian subduction zone (von Huene and Kulm, 1973) the Nankai Trough (Moore and Karig, 1976; Karig and Kagami, 1983) and the observed escape of fluids at the base of the Barbados (Moore, Biju-Duval 1983) and the Guatemalan margin (Aubouin, von Huene, et al, 1982) are direct observations of dewatering. Estimates of the rate of dewatering based on indirect techniques have been made by several authors(e.g. Karig, in press). The estimates suggest that porosities of 20-30% can occur 3 km deep in accreted sediment and thus about 20-30% of the interstitial water (originally 50%) have been lost. Once the sediment passes down the subduction zone to where its depth exceeds 3 km, the rise in temperature triggers a second source of water, that from the mineral dehydration reactions that occur with rising temperature (c.f. Miyashiro, 1978).

Thermodynamic constraints require dehydration of oceanic crust as it is subducted into higher temperature environments (Anderson et al, 1978). Hydrous minerals are converted to less-hydrous minerals in a series of transformations when temperature exceeds about 85,C, or the temperature where smectite converts to illite and other less hydrous clay minerals. The smectite-illite transition in sediment produces about 15% water (Bruce, 1973). A theoretical estimate of the amount of water produced by metamorphic dehydration is beyond the scope of this study and involves the study of diagenesis and metamorphism of at least three major elements; 1) the accreted and subducted sediment, 2) the upper basaltic ocean crust, and 3) the lower ultramafic ocean crust.

Surface venting of pore fluids (connate water) captured in subducted rock is being recognized with increasing frequency. Water derived from mineral dehydration, the so called "metamorphic water", was found in springs of the California Coast Ranges prior to a recognition of subduction (White, 1957) and is now found more frequently (White et al, 1973). Such water is found in land areas where the subducted crustal source is about 15-20 km deep. Further toward the trench, along the Washington-British Columbian continental shelf, large diapiric structures cored by mobilized shale are found where the oceanic crust is 8.5 km deep (Snavely and Wagner, 1981). West of the Mariana Arc, diapirs of serpentinized plutonic rock, is thought to signify venting of fluid charged rock from the subduction zone (Hussong and Fryer, 1983), at a depth of about 9 km (La Traille and Hussong, 1980). Mud diapers have been recognized by swath mapping in the Barbados forearc, high on the lower slope (Biju-Duval et al, 1982). These are but some examples of the venting fluids from deeper parts of a subduction zone.

Workers in several disciplines of the earth sciences have recognized the production of fluids due to metamorphism and their importance in subduction zones (c.f. Anderson et al, 1977 for thermal consequences; Barnes et al, 1978, metamorphic components in ground water; Fyfe and Kerrich, 1983, metamorphic consequences; Ringwood, 1977, magma generation). However, the difficult to construct quantitative models that account for all of the thermal-metamorphic processes in a modern subduction zone are not yet forthcoming. The depth of dehydration can be approximated by plotting the isotherms of the initial temperatures of common hydration reactions on existing models (fig. 11). Some dehydration temperatures of minerals common to different parts of the ocean crust are: 1) smectite to less hydrous clays + water, less than 50,C, 2) smectite to zeolites + water, 85 to 100,C, 3) zeolite to the greenshist assemblage of minerals + water, 200 to 300,C, 4) greenshist to the amphibolite assemblage of minerals, 350-550,C. The 550,C isotherm on the 20 m.y. model of Anderson et al (1978) reaches about 140 km, and the 250, isotherm of Bird (1978) for the Mariana and Tonga Trenches reaches about 125 km. Thus the latter phases of hydration may continue into the zone of partial melting according to some existing temperature models. The availability of water to provide overpressured conditions can extend far down a subduction zone under the proper circumstances but that possibility has not been fully tested.

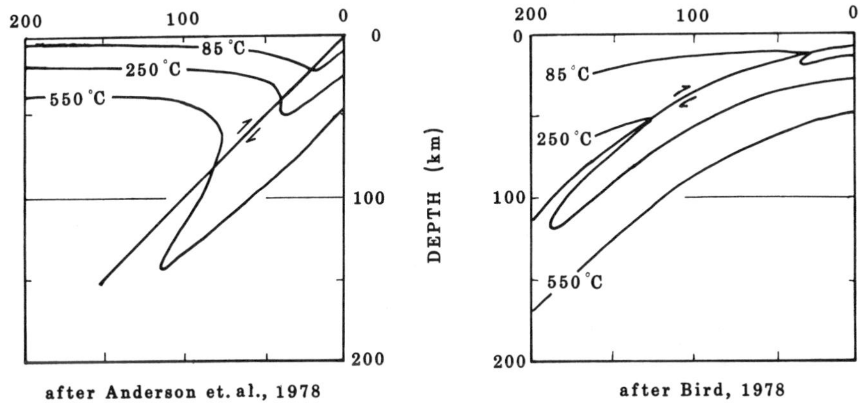

Figure 11. Two thermal models of subduction zones: a generalized model (Anderson et al, 1978) and a model of the Tonga and Mariana Arcs (Bird, 1978). The isotherms shown correspond to the dehydration temperatures from clay to zeolite (85,C), zeolite to greenschist (250,C), and greenschist to amphibolite (550,C). The differences in these two models probably reflects as much the state of knowledge as the differences between subduction zones.

CONCLUSIONS

Horizontal compressional deformation is confined principally to the slope of the trench along several modern convergent margins. Much compressional structure landward of the trench slope was formed previously and does not necessarily reflect the present stress field.

Older seismic reflection records that showed structure at only shallow depths and were displayed at high (greater than 5x) vertical exaggeration provided minimal constraints on compressional or tensional structure. The sharper resolution of modern multichannel records clearly shows strong compressional deformation across the lower slope, much vertical tectonics across some upper slopes, and and an absence of thrust faulting across forearc basins. Some volcanic arcs contain principally non-compressional structure. In the Aleutian Arc, the orientation and prevalence of only extensional structure is consistant with the directions of stress axis based on the study of secondary volcanic vents and dike patterns but the principal stress parallels the tensional rather than compressional direction.

Limited transmittal of stress through the subduction zone as suggested by the relatively narrow zone of compressional deformation. Although decoupling must result from globally varying conditions, such as configuration of the subduction zone, buoyancy contrasts between the upper and the lower plates, and absolute motion of the upper plate with respect to the trench, an equal decoupling can result from local causes. One local condition that is potentially present in any deep submarine environment, is the increase in pore fluid pressure. If sediment passing through a trench contains interstitial water which does not fully drain during accretion, it will attain greater than hydrostatic pressure and thus weaken the coupling across a subduction zone. As subducting sediment and ocean crust are carried across isotherms above 85,C, the dehydration accompanying metamorphism provides a continuous supply of water. When that water is trapped, pore pressure can approach lithostatic values.

The observed strain suggests that the stress from horizontal compression in the subduction zone can drop to low values a few tens of kilometers away from the zone. An explanation for a rapid stress gradient drop is the low strength of materials adjacent to the subduction zone which cannot transmit high levels of stress.

REFERENCES

Anderson, Roger N., DeLong, Stephen E., and W.M. Schwartz, 1977, Thermal model for subduction with dehydration in the downgoing slab: J. Geology, v. 86, p. 731-739.

Aoki, Y. and T. Tamano, (1983) Detail structure of the inner trench slope of the Nankai Trough, from migrated seismic sections: Amer. Assoc. Pet. Geol. Memoir 34, in press.

Aubouin, Jean, Roland von Huene, Miriam Baltuck, Robert Arnott, Jacques Bourgois, Mark Filewicz, Roger Helm, Keith Kvenvolden, Barry Lienert, Tom McDonald, Kristin McDougall, Yujiro Ogawa, Elliot Taylor, Barbara Winsborough, 1982, Leg 84 of the Deep Sea Drilling Project, Subduction without accretion: Middle America Trench off Guatemela: Nature v. 297, p. 458-460.

Aubouin, J., von Huene, R.E., Azema, J., Coulbourn, W., Cowan, D., Curiale, J., Dengo, C., Faas, R., Harrison, W., Hesse, R., Ladd, J., Muzylov, N., Shiki, T., Thompson, P., Westberg, J., 1982: Initial Reports of the Deep Sea Drilling Project, 67, Washington (U.S. Govt. Printing Office).

Bally, A.W., 1981, Thoughts on the tectonics of folded belts: in Thrust and Nappe Tectonics, Geol. Soc. London Spec. Pub no. 9, p. 13-32.

Barnes, Ivan, Irwin, W.P. and D.E. White, 1978, Global distribution of carbon dioxide discharges and major zones of seismicity: USGS Water Resources Investigations 78-39, 12 p.

Biju-Duval, B., LeQuellec, P., Mascle, A., Renard, V., and P. Valery, 1982, Multibeam bathymetric survey and high resolution seismic investigations on the Barbados Ridge complex (Eastern Caribbean): A key to knowledge and interpretation of an accretionary wedge: Tectonophysics, v. 86, p. 275-304.

Bird, Peter, 1978, Stress and temperature is subduction shear zones: Tonga and Mariana: Geop. J.R. Astr. Soc.,v. 55, p. 411-434.

Bostrom, R.C., 1971, Westward displacement of the lithosphere: Nature, v. 234, p. 256-538.

Brown, L., Klein, J., Middleton, R., Pavich, M., Sacks, I., Tera, F., 1982, Beryllium-10 as a geochemical and geophysical probe. Carnegie Institute Washington Yearbook, v. 81, p. 464-470.

Bruce, C.H., 1973. Pressured shale and related sediment deformation: mechanism for development of regional comtemporaneous faults. AAPG Bull. v. 57, no. 5, p. 878-886.

Bruns, Terry R., 1983, Model for the origin of the Yakutat block, an accreting terrane in the northern Gulf of Alaska: Geology, v. 11, p. 718-721.

Bruns, Terry R., Lewis, S.D., Blackman, D.K., von Huene, R., Ladd, J.W., 1983, Structure and Tectonics of the Shumagin Continental Margin, Western Gulf of Alaska, EOS, v. 64, no. 45, 870 p. (abstract).

Bruns, Terry R., and R.E. von Huene, 1977, Sedimentary basins on the Shumagin shelf, western Gulf of Alaska: 1977 Offshore Technology Conference Proceedings, v. 1, p. 41-50.

Carson, Bobb, von Huene, Roland, Arthur, Michael, 1982, Small-scale deformation structures and physical properties related to convergence in Japan Trench slope sediments: Tectonics, v. 1, no. 3, p. 277-302.

Davis, D., Suppe, J., and F.A. Dahlen, 1983, mechanics of fold-and-thrust belts and accretionary wedges: Jour. Geop. Research, v. 88, p. 1153-1172.

Fisher, M.A., and R. von Huene, 1980, Structure of Upper Cenozoic Strata beneath Kodiak Shelf, Alaska: Amer. Assoc. Petrol. Geol. Bull., v. 64, no. 7, July, p. 1014-1033.

Fisher, M.A., 1979, Structure and tectonic setting of the continental shelf southwest of Kodiak Island, Alaska: American Asociation of Petroleum Geologist Bulletin, v. 63, p. 301-310.

Fyfe, W.S., Kerrich, R., 1983, Fluid release and mineralization associated with thrusting, <u>in</u>

Extended abstracts, 4th International Symposium on water-rock interaction, 29 August - 3 September 1983. Institute for Thermal Spring Research, Okayama University, Misasa, Japan, p. 150-151.

Honza, E., Kagami, H., and N. Nasu, 1977, Neogene geological history of the Tohoku Island Arc System: Jour. Oceanog. Soc. of Japan, v. 33, no. 6, p. 297-310.

Hubbert, M.K., Rubey, W.W., 1959, Role of fluid pressure in mechanics of overthrust faulting: Geol. Soc. of Amer. Bull., v. 70, p. 115-166.

Hussong, Donald M., Fryer, Patricia, 1983, Back-Arc Seamounts and the Sea MARC II Seafloor Mapping System, EOS transactions, AGU, v. 64, no. 45, p. 627-632.

Hussong, Donald M., and Fryer, Patricia, 1982, Structure and tectonics of the Mariana arc and forearc: drillsite selection surveys, in initial reports of the DSDP, National Science Foundation, v. 60, p. 35-55.

Hussong, Donald M., Uyeda, Seiya, 1982, Tectonic processes and the history of the Mariana arc: a synthesis of the result of DSDP Leg 60, in Initial reports of the DSDP, National Science Foundation, v. LX, p. 909-929.

Hussong, D.M., Wipperman, L.K., 1981, Vertical movement and tectonic erosion of the continental wall of the Peru-Chile Trench near 11,30'S latitude. In Nasca Plate: Crustal formation and Andean Convergence. GSA Memoir 154, p. 509-524.

Ida, Y., 1978, Oceanic crust in the dynamics of plate motion and back-arc spreading: J. Phys. Earth, 26 (Suppl.), p. 355-568.

Ishiwada, Y., and K. Ogawa, 1976, Petroleum geology of offshore areas around the Japanese islands, UN-ESCAP, CCOP Tech. Bull., v. 10, p. 23-34.

Jordan, T., Isacks, B., Allmendinger, R., Brewer, J., Ramos, V., Ando, C., 1983, Andean tectonics related to geometry of subducted Nazca Plate: Geol. Soc. Am. Bull., v. 94, no. 3, p. 341-361.

Karig, Daniel E., in press, The framework of deformation in the Nankai Trough.

Karig, Daniel E., Kagami, Hideo, 1983, Varied responses to subduction in Nankai Trough and Japan Trench forearcs: Nature, v. 304, no. 5922, p. 148-151.

Karig, Daniel E., Kagami, Hideo, et al, 1983, Varied responses to subduction in Nankai Trough and Japan Trench forearcs: Nature, v. 304, no. 5922, p. 148-151.

Karig, D.E., Moore, G.F., Curray, J.R., and Lawrence, M.B., 1980, Morphology and Shallow Structure of the Lower Trench Slope off Nias Island, Sunda Arc, _in_ the tectonic and geologic evolution of southeast Asia seas and islands, AGU geophysical nomograph, 23, p. 179-208.

Keller, G., 1980, Benthic foraminifers and paleobathymetry of the Japan Trench area, Leg 57, Deep Sea Drilling Project, in Scientific Party, Initial Reports of the Deep Sea Drilling Project, 56, 57, pt. 2, Washington (U.S. Govt. Printing Office), p. 835-866.

Keller, G., von Huene, R.E., McDougall, K., Bruns, T.R., in press, Paleoclimatic evidence for Cenozoic migration of Alaskan terrances.

Kulm, L.D., Prince, R.A., French, W., Johnson, S., and A. Masias, 1981, Crustal structure and tectonics of the central Peru continental margin and trench, GSA Memoir 154, p. 445-468.

Kulm, L.D., von Huene, R.E., Duncan, J.R., Ingle, J.C., Kling, S.A., Musich, L.F., Piper, D.J.W., Pratt, R.M., Schroder, H.J., Weser, O., and Wise, S.W., 1973, Initial reports of the Deep Sea Drilling Project: v. 18, Washington, D.C., U.S. Government Printing Office, 1077 p.

Ladd, John W., in press, Seismic stratigraphy of the continental shelf offshore Guatemala: implications for vertical tectonics related to subduction, DSDP, v. 84.

Ladd, J.W., Ibrahim, A.K., McMillen, V.J., Latham, G.V., and R.E. von Huene, 1982, Interpretation of seismic reflection data of the middle America Trench offshore Guatemela, _in_ von Huene, R., Aubouin, J., et al, 1982, Initial Reports of the Deep Sea Drilling Project, v. 67, Washington

(U.S.G.P.O.) p.
Lehner, P., Doust, H., Bakker, G., Allenbach, P., Gueneau, J., 1983, Active margins, Part 2; Tonga Trench, Profiles P-1200 and G-150, in Tectonics of compressional provinces/strike-slip tectonics, v. 3.

Marsh, B.D., 1980, Geology and petrology of northern Atka, Aleutian Islands, Alaska: GSA abstracts w/Programs, v. 12, no. 7, 476 p.

McCarthy, Jill, and David W. Scholl, in press: Mechanisms of subduction accretion along the central Aleutian Trench, GSA Bull.

McKenzie, D.E., 1969, Speculations of the consequences and causes of plate motions: Geop. J. Roy. Astron. Soc., v. 18, p. 1-32.

Mclean, H., et al, 1980, Geology of Amlia Island, Aleutian Islands, Alaska: Cordilleran Meeting, GSA, Abstracts w/Programs, v. 12, no. 32, 119 p.

Miyashiro, A., 1978, Metamorphism and metamorphic belts: George Allen and Unwin, Publishers, London, 481 p.

Moore, Casey J., and Bernard Biju-Duval, in press, Tectonic Synthesis Leg 78A DSDP: Structural evolution of offscraped and underthrust sediment northern Barbadas Ridge complex.

Moore, Casey J., and Bernard Biju-Duval, et al, 1983, Offscraping and underthrusting of sediment at the deformation front of the Barbados Ridge: results from Leg 78A, DSDP, GSA Bull. v. 93, p. 1065-1077.

Moore, Casey J., and D.E. Karig, 1976, Sedimentology, structural geology, and tectonics of the Shikoku subduction zone: Geol. Soc. of Am. Bull., v. 87, p. 1259-1268.

Moore, G.F., and D.E. Karig, 1976, Development of sedimentary basins on the lower trench slope: Geology, v. 4, p. 693-697.

Moore, C.J., Biju-Duval, B., et al, 1981, Near Barbados Ridge scraping off, subduction scrutinized: Geotimes, v. 26, no. 10, p. 24-26.

Mrozowski, C., and D.E. Hayes, 1980, A seismic reflection study of faulting in the Mariana forearc, in Hayes, D.E. (Ed.), The Tectonic and

Geologic Evolution of Southeast Asian Seas and Islands: Washington, (AGU), p. 223-234.
Nakamura, Kazuaki, Plafker, George, Jacob, K.H., and J.H. Davies, 1980, A tectonic stress trajectory map of Alaska using information from volcanoes and faults: Bull. of the Earthquake Research Inst., v. 55, p. 89-100.
Nakamura, Kauaki, and Seiya Uyeda, 1980, Stress gradient in arc-back arc regions and plate subduction: Jour. Geop. Resear., v. 85, no. B11, p. 6419-6428.
Nasu, N., R. von Huene, Y. Ishiwada, M. Langseth, T. Bruns and E. Honza, 1980, Interpretation of multichannel seismic reflection data, legs 56 and 57, Japan Trench transect, Deep Sea Drilling Project, in Scientific Party, Initial Reports of the Deep Sea Drilling Project, 56, 57, Washington (U.S. Govt. Printing Office), p. 489-504.
Scheidegger, K.F., Corliss, J.B., Jezek, P.A., and D. Ninkovich, 1980, Compositions of deep-sea ash layers derived from north pacific volcanic arcs: variations in time and space: J. of Volcanology and Geothermal Res., v. 7, p. 107-137.
Scholl, David W., Marlow, Michael S., Buffington, Edwin C., 1975, Summit Basins of Aleutian Ridge, North Pacific: AAPG Bull, v. 5f9, no. 5, p. 799-816.
Scholl, D.W., Vallier, T.L., Stevenson, A.J., 1983, Arc, forearc, and trench sedimentation and tectonics; Amlia corridor of the Aleutian Ridge, in Studies in Continental Geology, AAPG Memoir 34, Watkin and Drake, ed., p. 413-239.
Scientific Party, 1980. Initial Reports of the Deep Sea Drilling Project, 56, 57: Washington (U.S. Govt. Printing Office).
Seely, D.R., 1979, The evolution of structural highs bordering major forearc basins, in J.S. Watkins, L. Montadert and P.W. Dickerson, eds., Geological and Geophysical Investigations of Continental margins: Amer. Assoc. of Petroleum Geologists Memoir 29, p. 245-260.
Seely, D.R., Vail, P.R., and G.G. Walton, 1974, Trench slope model, in Geology of Continental Margins,

Burk, C.A., and C.L. Drake, eds., Springer-Verlag, New York, p. 261-283.

Shipley, T.H., Ladd, J.W., Buffler, R.T., and J.S. Watkins, 1982, Tectonic processes along the Middle America Trench inner slope; in J.K. Leggett, ed., Trench-Forearc Geology: Geol. Soc. London Special Pub. 10, p. 95-106.

Shipley, T.H., McMillen, K.J., Watkins, J.S., Moore, J.C., Sandoval-Ochao, H., and J.L. Worzel, 1980, Continental margin and lower slope structures of the Middle America Trench near Acapulco, Mexico: Marine Geology 35, p. 65-82.

Sleep, Norman, and M.N. Toksoz, 1973, Evolution of marginal basins: Nature, v. 233, p. 548-550.

Snaveley, P.D. Jr., Wagner, H.C., 1981, Geologic cross section across the continental margin off Cape Flattery, Washington, and Vancouver Island, British Columbia. USGS, Open-File Report, p. 81-978.

Suarez,G., Molnar, P., and B.C. kBurchfield, 1983, Seismicity, fault plane solutions, depth of faulting, and active tectonics of the Andes of Peru, Ecuador, and southern Columbia: Jour. Geophys. Res., v. 88, B12, p. 10, 403-428.

Uyeda, Seiya, 1982, Subduction zones: an introduction to comparative subductology: Tectonophysics, v. 81, p. 133-159.

Vallier, Tracy L., Scholl, D.W., Stevenson, A.J., and J. McCarthy, 1983, Contrasting tectonic responses to subduction processes in the Aleutian and Tonga Island Arcs, in OJI international seminar on the formation of ocean margins: Ocean Research Inst., Univ. Tokyo (Abstract).

von Huene, Roland, and Homa Lee, 1983, The possible significance of pore fluid pressures in subduction zones, in Studies in continental geology, AAPG Memoir 34, p. 781-791, Watkins and Drake, ed.

von Huene, Roland, Langseth, Marcus, Nasu, Noriyuki, Okada, Hakuyu, 1982, A summary of cenozoic tectonic history along the IPOD Japan Trench transect: GSA Bull., v. 98, p. 829-846.

von Huene, R., Moore, G.M., Moore, C.J., and C.D.

Stephens, 1979, Cross section, Alaska Peninsula - Kodiak Island - Aleutian Trench: Summary. GSA Bull., Part 1, v. 90, May, p. 427-430.
von Huene, R., Moore, G.W., and J.C. Moore, 1979, Cross section, Alaska Peninsula - Kodiak Island - Aleutian Trench: Geological Society of America Map and Chart Series, MC-28A, scale 1:250,000.
von Huene, R., and L.D. Kulm, 1973, Tectonic summary of Leg 18: Deep Sea Drilling Project Initial Repts., v. 18, p. 961-976.
Watkins, J.S., Moore, J.C., et al, 1981, Accretion, underplating, subduction, and tectonic evolution, Middle America Trench, southern Mexico: results from DSDP Leg 66, Oceanologica Acta, Supplement au volume 4, p. 213-223.
White, D.E., 1957, Magmatic, connate and metamorphic waters: GSA Bull., v. 68, p. 1659-1682.
White, D.E., Barnes, Ivan, and J.R. O'Neil, 1973, Thermal and mineral waters of nonmetric origin, California Coast Ranges: GSA Bull., v. 84, p. 547-560.

THE MEDITERRANEAN SEAS

LE PICHON, XAVIER
Laboratoire de Géodynamique, Université P. et M. Curie,
4 Pl. Jussieu, T.15, ler ét. 75230 Paris Cedex 05, France.

ABSTRACT

A key fact, concerning the Mediterranean seas, is the existence of the Calabrian and Hellenic intermediate seismic zones which can only be explained by subduction of oceanic lithosphere in the last 20My. The minimum amounts of subduction are given by the lengths of the seismic zones and correspond to 650 km and 500 km respectively. Subduction is still active today and has occurred at an average rate of about 4 cm yr^{-1}. The Balearic and North Algerian basins have an oceanic lithosphere which was formed by accretion 24-18 My ago. The Tyrrhenian basin has small portions of oceanic lithosphere formed in the last 12-7 My. Extremely thinned continental crust is found next to the oceanic crust in these basins. The amount of extension in the West Mediterranean basins is of the right amount to account for the length of the Calabrian intermediate seismic zone. In the same way, the amount of extension in the still continental Aegean sea is of the right amount to account for the length of the Hellenic intermediate seismic zone. The Eastern Mediterranean basins are covered by 4 to 12 km of sediments in the central basins and more on the Mediterranean ridge. The crust is now demonstrated to be of oceanic character in the central basins although controversy still exists about the Mediterranean ridge. The East Mediterranean lithosphere was most probably formed during Upper Mesozoic time after a phase of rifting which began in Jurassic. About half the surface of this Mesozoic basin has now disappeared by subduction. The Mediterranean basins were deeper than now at the beginning of the Messinian dessication phase. As a result, unloading of water must have concentrated tectonic stresses along the margins, stresses

Proceedings of the 27th International
Geological Congress, Volume 23, pp. 189–222
ORIGIN AND HISTORY OF MARGINAL AND INLAND SEAS

which might have locally induced overthrusting of the margins over the basin.

INTRODUCTION

Our knowledge of the structure of the Mediterranean seas has greatly increased in the last few years. This progress in knowledge has been recorded in a series of extensive collections of review papers such as Biju-Duval and Montadert (1977), Nairn et al. (1977), Hsü, Montadert et al. (1978), Wezel (1981) for the Mediterranean in general and Le Pichon et al. (1982a) for the Hellenic arc and trench. A most instructive paper is the one written by Hsü (1978). Actually, the conclusions he made at the time are still valid, in my opinion. More recent synthesis papers include Finetti (1982) for the Ionian sea, with and extensive bibliography, Rehault et al. (1984) for the Western Mediterranean basin and Westphal et al. (1984) for the paleomagnetics of the Mediterranean area.

Yet, in the last three years, important new data on the structure and evolution of the basins have been obtained. Seismic reflection and refraction measurements now precisely delineate the ocean crust limit in the Balearic basin and quantify the exact amount of thinning on the adjacent continental crust (Le Douaran et al., 1984). These data have been used by Burrus (1984) to propose a tectonic reconstruction of the Balearic basin. Seismic reflection and refraction measurements also give a good evaluation of the amount of crustal thinning in the Tyrrhenian basin (Steinmetz et al., 1983; Recq et al., 1984) and the oceanic crust limits can be tentatively delineated there (Moussat, 1983). A good quality seismic refraction line has been obtained in the Levantine basin (Makris et al., 1983). Another one has been shot between Egypt and Crete (Makris, 1981), although the detailed results are not yet published. We have a better knowledge of the configuration of the Tyrrhenian deep seismic zone (Gasparini et al., 1982). Additional western Mediterranean heat-flow measurements and a careful heat-flow synthesis have been published by Hutchinson (1983). Precise paleomagnetic results on the Neogene rotations of Sardinia (Edel et al., 1981) and of the Hellenic arc (Laj et al., 1982; Kissel et al., 1984) are available. Finally, a submersible investigation of the passive margins of the Ionian sea provides important data bearing on their evolution (Escarmed, 1982).

Thus, another extensive review is probably not justified at that time, and I try instead in this paper to evaluate the significance of these new results concerning the structure and evolution of the Mediterranean basins. I concentrate on two aspects of this evolution. First, I discuss the origin of the Mediterranean basins within the context of the Africa-Europe collision, following an approach elaborated on the basis of the recent evolution of the Aegean area (Le Pichon and Angelier, 1979,1981, Le Pichon, 1982). Secondly, I discuss the possible impact of the Messinian salinity crisis on the tectonics of the continental margins.

THE HELLENIC AND CALABRIAN INTERMEDIATE SEISMIC ZONES

A key fact concerning the Mediterranean basins is the existence of two zones of intermediate depth earthquakes, one dipping to the northeast below the Hellenic arc and the Aegean sea (Papazachos, 1973, 1976), the other dipping to the west-northwest below the Calabrian arc and the Tyrrhenian sea (Gasparini et al. 1982). The Hellenic intermediate seismic zone extends to a maximum depth of 150-175 km with a rather shallow dip. Hence, errors in localization, (probably up to 30-50 kilometers) make the precise determination of the shape of the zone difficult. Thus, Horvath et al. (1981) argue that this zone results from continental collision, not from oceanic lithosphere subduction. On the other hand, the Calabrian intermediate depth zone extends down to a depth close to 500 kilometers Consequently, although the scatter is also quite large, the zone appears to be better defined. It was previously thought that the zone is not continuous between 50 and 200 km and this was interpreted as resulting from a detached slab (e.g. Horvath et al., 1981). However, earthquakes have now been detected in this depth range and the zone is consequently continuous (Gasparini et al., 1982) with an average dip of about 50°, the deeper portion being somewhat flatter. The total length of the Hellinic intermediate seismic zone, counted along the direction of underthrusting given by surface earthquakes, is a maximum of 500 km (Le Pichon and Angelier, 1979). The total length of the Calabrian seismic zone is 650 kilometers. Both zones have awidth of 250-300 kilometers, taken perpendicular to the general direction of underthrusting although both tend to adopt a fan shape at depth.

The existence of earthquakes at depths larger than 80-100 km is difficult to explain by continental collision. Otherwise, one should find intermediate seismic zones in regions of rapid collision, as under the Himalayas. It cannot either be explained by the detachment of the lower portions of the lithosphere, which must be occurring in regions of continental crust stacking as in Tibet (Allègre et al., 1984) where it is not accompanied by intermediate earthquakes. Actually, the evidence available indicates that earthquakes only occur in the relatively cold, upper brittle portion of the lithosphere (e.g. Meissner and Strehlau, 1982). And no evidence exists to suggest that it is possible to insert at large depths the buoyant upper portion of a continental lithosphere. I conclude that both Mediterranean imtermediate depth seismic zones result from geologically rapid subduction of negatively buoyant lithosphere. The negative buoyancy implies that the crust of this lithosphere must be less than about 15 km thick (e.g. Le Pichon, 1982).

The lithosphere which has been subducted in both cases is former Eastern Mediterranean lithosphere. The igneous crust is indeed generally less than 15 kilometers, as we shall see later. But this reasoning implies that the large sedimentary cover does not partake in the subduction (Le Pichon and Angelier, 1981).

Molnar et al. (1979) show that the maximum possible length of a subducting slab is given approximately by v/A 10, where v is the rate of subduction and A the age of the lithosphere. Consequently a subducted lithosphere 100 M.y. old ceases to be seismogenic about 10 M.y. after subduction and a lithosphere 200 M.y. old about 20 M.y. after subduction. I conclude that both intermediate earthquake zones in the Mediterranean result from subduction, active during the last 10 to 20 M.y. at most, at rates of 60 to 30 mm yr^{-1}, of oceanic or quasi-oceanic lithosphere. This lithosphere must be at least 100 M.y. old, otherwise the rates of subduction implied become unreasonably high. I now investigate whether these inferences are borne out by independent geophysical and geological data.

STRUCTURE AND FORMATION OF THE BALEARIC BASIN

Figure 1 summarizes the main conclusions, based on geophysical and geological data, concerning the nature of the Western Mediterranean bsins. In the Balearic basin, the results of seismic investigations are clear (Le Douaran et al., 1984). The limits of the oceanic crust (Fig.1) have been precisely determined on the basis of thickness of igneous crust, velocity, character of basement reflection and slope of regional basement topography (Burrus, 1984). An important result is that, adjacent to the oceanic crust, zones of extremely thinned continental crust (less than 5 km in places!) exist. Similar thinning has been demonstrated elsewhere in the oceans (see for example Avedik et al., 1982). For most geodynamic purposes, this highly thinned continental crust is the exact equivalent of the oceanic crust and it is not possible to differentiate between both types. Note that the crust of the Balearic islands is also thinned; its structure is suggestive of a Rift type (Banda et al., 1980) as the igneous crust is less than 15 km thick and the Moho is at 20 to 25 km depth. Note also that the upper mantle velocity is slow, in agreement with the measured high heat flow (see later).

Paleomagnetic data indicate that Corsica has rotated anticlockwise by 30° with respect to Europe since Permian whereas the rotation angle is 60° for Sardinia (Westphal et al., 1984). Edel et al (1981) have shown further that 30° of the rotation of Sardinia occurred between 21 and 19 M.y. ago, at the beginning of Miocene.

Microstructural field analysis of Eocene directions of compression confirm that there has been a rotation of 30° of Sardinia since Eocene (Letouzey and Trémolières, 1980). This date of opening is confirmed by convergent geological data which are summarized for example by Rehault et al. (1984) or Burrus (1984). I only mention here the results of submersible investigations we made on the Provence margin which demonstrate without ambiguity that this margin began to subside in Upper Oligocene (Chattian) 30 to 25 M.y. ago (Estocade, 1978). The transition from continental rifting to oceanic accretion most probably occurred in Aquitanian, between 24 and 23 M.y. ago (Burrus, 1984) and the proces of accretion stopped 5 M.y. later during Burdigalian. Burrus (1984) has shown that it is then possible to interpret the magnetic lineations, first interpreted as oceanic by Bayer et al. (1973), as corresponding to anomalies 6 to 60 of La Brecque et al.(1977) and to obtain a reasonable reconstruction of the opening between Europe and the Corso-Sardinian block (figure 2).

In a paper which was rejected in 1972 as too speculative, we pointed out with J.P.Foucher that the isostatically corrected depth and the heat flow within the Balearic basin were compatible with a 25 M.y. old oceanic lithosphere. It is now demonstrated that the heat flow in marginal basins fits the standard oceanic lithosphere decrease with age whereas the isostatically corrected depth-age curve is about 1000 meters deeper than the oceanic one (Louden, 1980). The average heat flow corrected for sedimentation rate is 109 $mWm^{-2} \pm 15$ according to Huchinson, which corresponds to an age of 15 to 25 M.y., in excellent agreement with the age obtained above.

The average stratigraphic column corresponds to 2570 meters of water, 1050 meters of Plio Quaternary (density: 2000 Kg m^{-3}), 1580 meters of evaporites (2200 kg m^{-3}) and 2300 meters of pre-Messinian (2400 kg m^{-3}). Taking a density of 3200 kg m^{-3} for the asthenosphere, based on a 0°C density of 3350 kg m^{-3}, the isostatically corrected depth is 4730 meters. Louden (1980) has shown that the water depth w, in meters, in marginal basins vary approximately as

$$W = 3500 + 350 \quad A^{1/2}$$

where A is the age in My. With W = 4730, A = 12 My. Using 300 m instead of 350, as proposed by Hayes in still unpublished work, we have A = 17 My. Considering that the zero age depth is known to vary by several hundred meters, the age obtained is in agreement with previous conclusions. Consequently, the different geophysical arguments are compatible with an age of formation of the oceanic lithosphere of 24 to 19 M.y. ago, the structure being of the marginal basin type.

The reconstruction proposed by Burrus (B in figure 2) is the reconstruction immediately before accretion started. The total rotation adopted for Corsica is 14° about a pole near 44.2°N, 10°E and for Sardinia 30° about a pole near 43.1°N and 9.2°E. These rotations maintain continuity between Corsica and Sardinia although there is a 15° differential rotation which should have produced extension to the west of the straits of Bonifacio and compression to the east. The next question I discuss is whether the preceding rifting stage involved significant extension between Corsica-Sardinia and Europe.

The careful study of Le Douaran et al. (1984) shows that the continental crust is highly thinned to a thickness of about 5 kilometers, equivalent to the thickness of the accreted oceanic crust. This observation seems to apply to other continental margins as well. As pointed out by Le Pichon and Sibuet (1981), it implies that accretion starts where the continental crust has sufficiently thinned to reach the depth equivalent to the depth of a zero age oceanic lithosphere. Does this thinning conserve the volume of the continental crust or does it result from subcrustal erosion or mantle assimilation?

If continental crust is conserved, it is easy to restore the pre-stretching continental boundary. Le Douaran et al. show that, over four France continental margins profiles, the average amount of stretching stays close to 170 % although the width of the margin varies by a factor of four. This is because the decrease in thickness of crust is approximately linear over the whole width of the margin. This linear decrease seems to be characteristic of most margins as noted by Le Pichon and Sibuet (1981). In this case, let b_x be the maximum stretching ratio, the average stretching is $\bar{b} = 2b_x/(1 + b_x)$. In the Balearic basin, b_x is about 5(25/5 km) and thus $\bar{b}$ = 167%. The self-regulation mechanism which maintains this linear decrease in the thickness of the crust may be related to the stresses produced by lateral density changes (Le Pichon, 1982). In figure 1, I have used the 1.7 ratio to find the position of the edge of the unstretched continental crust. Figure 2 illustrates that the amount of relative displacement obtained for Sardinia during the rifting stage is 120 to 150 km whereas it is only one third of that for Corsica. Although this estimate is not precise, the total amount of stretching is about three times less for Corsica and Sardinia. To maintain continuity between Corsica and Sardinia would require an additional 30° of rotation for Sardinia and about 15° for Corsica. This is apparently excluded by paleomagnetic and tectonic observations. Then, the only possibility left, with this hypothesis of conversation of continental crust, is the one shown in figure 2, with a decoupling of Corsica and Sardinia along the straits of Bonifacio. Note that, in this solution, the Oligocene Sardinian Rift gets in the prolongation of the Corsica-Provence Rift. A small additional rotation of Sardinia could also put it in the prolongation of the Valencia Rift between the Baleares and Spain.

One can instead assume that there was no stretching, only thinning by subcrustal erosion or assimilation. This is contrary to the evidence of significant extension in the upper brittle crust (Le Douaran et al., 1984). However, many researchers doubt that the amount of extension is as large as required by constant volume thinning. If they are correct, the positions of Corsica and Sardinia prior to rifting would lie somewhere between positions A and B in figure 2. In this paper, I assume, following Le Pichon and Sibuet (1981), that continental crust is conserved, although this problem certainly cannot be considered to be solved today. I will show that the results of such an assumption lead to a reasonable agreement between the amount of Calabrian subduction and the amount of total West Mediterranean opening.

STRUCTURE AND FORMATION OF NORTH ALGERIAN AND ALBORAN BASINS

There has been much less detailed seismic reflexion and refraction work in these basins than in the Balearic basin. However, we know that the crust is only 5 km thick in the North Algerian basin under a thickness of 4 km of sediments and the Moho depth is about 12 km (Hinz, 1973). The structure then, appears to be the same as in the center of the Balearic basin and this is confirmed by the similarity in the sedimentary filling, as deduced from seismic reflexion (Rehault et al., 1984). Well developed magnetic lineations exist north of Algeria between the longitudes of Baleares and Sardinia. They show a symmetry and have been interpreted as indicating an ESE-WNW opening there (Bayer et al., 1973, see figure 1). Further west, the magnetic lineations have NNW-SSE orientation and are probably best interpreted as related to fracture zones although the width of the basin is so small that one cannot be sure that the anomalies are of the sea-floor spreading type.

The Alboran basin, on the other hand, has a somewhat thicker crust, the Moho being at a depth of about 20 km (Hatzfeld, 1978). The magnetic anomalies there are probably best explained by a combination of tectonics and volcanism (Galdèano et al. 1974).

Although the Alborin basin is probably not oceanic, its thinned continental crust was produced by extension some time prior to Burdigalian, most probably at the same age as the whole Balearic and North Algerian basins (Rehault et al., 1984). The only possible direction of extension for the formation of the basin is NE-SW (Le Pichon et al., 1972), based on the direction of the Alboran sea ridge and on the strikes and senses of motion of the strike-slip faults which were active on land to the south in Lower Miocene (Leblanc and Olivier, 1984).

However, a complication exists in this area because there is strong evidence for the west-south-west motion of an Alboran block in Lower Miocene time (Andrieux et al., 1971). The northward limit was along a right lateral North

Betic fault along which 200 km of motion may have occurred (Leblanc and Olivier, 1984). Why did this block move to the west and how is this motion compatible with the opening of the North Algerian basin to the south-southeast? At this time, the motion of Africa with respect to Europe was to the north or slightly east of north (Olivet et al., 1984). The closest contact between Iberia and Africa did not occur between Morocco and Spain but between western Algeria and Iberia. It is possible that collision induced expulsion to the west of the Alboran block which was situated in the region of closest contact (Tapponier, 1977).

STRUCTURE AND ORIGIN OF TYRRHENIAN BASIN

The structure of the Tyrrhenian sea is well known although the exact limit of the oceanic-continental crust boundary is not established. The Moho is very shallow (10 to 11 km) in the deepest plains (Finetti and Morelli, 1973), where the crust is about 5 kilometers thick or even less (Steimetz et al. 1983; Recq et al., 1984). In the thinnest crust portion, the presumably continental 6 km s^{-1} layer disappears over a width of about 100 km along a profile at the latitude of 40°N (Recq et al., 1984). Using this criterion, the eastern limit of the Sardinian thinned continental crust is along the Central Fault, slightly west of 12°E. However, detailed seismic reflexion studies indicate the presence of tilted blocks over a significant portion of this thinned crust area (Moussat, 1983). Moussat consequently proposes that oceanic crust produced by accretion be limited to three large tholeiitic volcanoes in the western central plain (Magnaghi and Vavilov) and in the Eastern plain (Marsili). These seamounts are uncontaminated oceanic tholeiites erupted between 7 and 3 Ma (Barberi et al., 1978). Thus, the interpretation of Moussat is a minimum one; the absence of the "continental" 6 km sec^{-1} layer over an area significantly larger than just these seamounts suggests that the oceanic crust may extend beyond these seamounts too, as suggested in figure 1.

The sedimentation corrected heat flow in the presumably continental portion of the Western Central plain is 145 $mWm^{-2} \pm 13$ (Hutchinson, 1983) which would correspond to an age of 9 to 13 M.y. for an oceanic crust and slightly more for a thinned continental crust. In the presumably oceanic Eastern portion of the plain, the sedimentation corrected heat flow is 161 $mWm^{-2} \pm 14$ (Hutchinson, 1983) which corresponds to an age of 7 to 10 M.y. for an oceanic crust. The isostatically corrected water depth is about 3600 to 3650 meters in the Eastern plain where the sediment thickness is small (for example near site 373 of DSDP). This depth would indicate a 2500 to 2800 meters zero age depth for a thermal subsidence of the oceanic type if the age is 7 to 10 M.y. Such a zero age depth is small for a marginal basin (Louden, 1980) which suggests that the oceanic area is still too small to reach its equilibrium depth. In addition, geologic evidence suggests that the Central Basin was already deep in Messinian time whereas subsidence to the east is younger (Fabbri and Curzi, 1979).

The magnetic anomalies in the Central and Eastern plains are lineated in a N-S to NNE-SSW direction (AGIP, 1981). A symmetry may exist in the Eastern plain (Selli, 1970) but an interpretation in terms of accretion is difficult to establish in a convincing way. Gravity anomalies confirm that the Tyrrhenian, as well as all the Western Mediterranean basins, are in isostatic equilibrium (Finetti and Morelli, 1973).

The evidence for distension either on the margins or in the basin is overwhelming and the direction of distension is well established to be N 120 E (Moussat, 1983). The age at which the basins first reach a significant depth may be 7 to 12 M.y. in the west and 5 to 8 in the southeast (Hutchinson, 1983). Extension is still active (Moussat, 1983). The amount of distension, assuming conservation of continental crust and using the limits of oceanic crust given in figure 1, is about 350-400 km in a direction of N 120° E in the axis of the Tyrrhenian basin. This amount of distension rapidly decreases to the northeast toward the Apennines and to the southwest toward Sicily. If, on the other hand, it is assumed that continental crust has been thinned but not stretched, the amount of extension is limited to the oceanic crust area which is about 100-150 km.

Paleomagnetics indicate no significant rotation of autochtonous Apulia since Upper Cretaceous time although the allochtonous Apennines (based on measurements in Umbria) may have rotated anticlockwise by about 30° sometime in Lower Miocene (Westphal et al., 1984). On the other hand, Calabria has a large anticlockwise rotation which exceeds the rotation of Sardinia by about 50°. It is thusprobable that Calabria rotated anticlockwise during the opening of the Tyrrhenian basin.

CALABRIAN SUBDUCTION AND WEST MEDITERRANEAN BASINS EXTENSIONS

We can now compare the age and amount of subduction indicated by the intermediate seismic Calabrian zone with the age and amount of extension in the Western Mediterranean basins. Both should be comparable because the maximum amount of shortening between Europe and Africa in the last 20 M.y. is less than 140 km in a N-S direction at a longitude between Corsica en Calabria (Olivet et al., 1984). Thus, to obtain 650 km of subduction in a direction of 305-310°, it is necessary to have about 600 km of extension in a direction of N 120°E.

Let us first consider the minimum amount of extension, which corresponds to oceanic crust. It is 100-150 km in the last 12-8 M.y. for the Tyrrhenian and 200-250 km, at the level of Sardinia in a N 120° E direction, in the last 23-19 M.y., for the Balearic basin. The total amount (300-400 km) is still too small by at least 250 km, especially if we consider that the subducted lithosphere was unlikely to be more than 120-160 M.y. old (Biju-Duval et al., 1977). Using Molnar et al. (1979) observations, a slab of lithosphere 160 M.y. old will cease to be seismogenic a minimum of 11 M.y. and a maximum of 25 M.y. after subduction. Thus, it is quite possible that the part of the lithosphere subducted in Lowermost Miocene may not be seismogenic any more.

I now use the estimates of extension based on oceanic accretion plus continental crust conservation during stretching. One obtains 350-400 km in the last 18 M.y. (since Burdigalian) for the Tyrrhenian which is still insufficient. The amount of extension is about right if one includes the 200-250 km of oceanic accretion in the Balearic basin between 23-19 M.y. but becomes too large by adding the additional 200-250 km of Balearic Oligocene continental stretching (between 30 and 23 M.y.). As seen above, a lithosphere subducted more than 25 M.y. ago most probably is not seismogenic any more.

I conclude that the evidence indicates that the present 650 km long intermediate seismic Calabrian zone is related to the subduction of an oceanic lithosphere, subductio which started at the latest 23 M.y. ago at the beginning of Miocene and more probably sometime in Oligocene. The subduction is directly related to the opening of the Balearic and Tyrrhenian basins which occurred in succession and involved about 800 km of total extension, half being produced by oceanic accretion and half by continental crust stretching but the lithosphere subducted in late Oligocene is not seismogenic any more.

An examination of the distribution of the amount and directions of extension in figure 1 indicates that the subduction was progressively funnelled to the southeast in an increasingly narrower basin. Consequently, subduction was terminated by collision at both its western and northern extremities while it continued and is apparently still continuing along the axis of the basin. It is well known that collision did indeed occur between Corsica and Apulia and between the Kabylies and Africa and reached its climax in Middle Miocene when the African continental basement was underthrust under the Kabylies (e.g. Durand-Delga, 1981). It is most probably this collision which led to the jump of the extension zone to the east of Sardinia, over the future Tyrrhenian basin. In the same way, collision has occurred on both sides of the subduction zone, in the Apennines and in Sicily, during the opening of the Tyrrhenian sea.

In this context, coming back to the Alboran block westward migration in Oligocene-Lower Miocene, it is quite possible that this migration triggered the active subduction below the eastern margin of the Iberian block. The exact localization of the extensial zone was probably in part governed by the preexisting Oligocene Rifts which extended to the north-north-east (Le Pichon et al., 1971). Note that the pattern of deformation in the whole Western Mediterranean has changed and is now dominated by continental collision (Bousquet and Philip, 1981).

THE HELLENIC SUBDUCTION ZONE: MIRROR IMAGE OF THE CALABRIAN ONE

I have just shown that the Eastern Mediterranean extended 600 to 800 km further to the northwest 25 to 30 M.y. ago to the north of autochtonous Sicily and Africa. It also extended up to 300 to 500 km to the northeast in the Aegean area. Its northern boundary to the east lay approximately along 38° of latitude in Oligocene time (Le Pichon and Angelier, 1979,1981). Paleomagnetic data confirm that subduction was accompanied by a 30° clockwise rotation of Peloponnesus since Uppermost Miocene (Laj et al., 1982). An earlier 30° clockwise rotation during Lower Miocene has been put into evidence in the Hellenides (Kissel et al., 1984).

The 300 to 500 km length of the Hellenic intermediate seismic zone and its NE direction of underthrusting cannot be explained by the 140-170 km of north-south Africa-Europe shortening in the last 20 M.y. Le Pichon and Angelier (1979) showed that this subduction implied the existence of a maximum simultaneous 350 km extension in the Aegean area, extension which resulted in continental stretching and subsidence (Hsü, 1978; McKenzie, 1978). The extension began about 10-13 M.y. ago (Le Pichon and Angelier, 1979). For example, the deeply subsiding North Aegean trough (Le Pichon et al., 1984) and the Sea of Crete (Hsü, 1978) were already formed in Messinian. But the rates of expansion and subduction were apparently accelerated 5 M.y. ago, based in particular on paleomagnetic evidence (Laj et al., 1982). However, the Aegean sea is still in an early continental stretching stage and may be about at the same stage as the Tyrrhenian basin 10 M.y. ago.

The average rate of subduction along the Calabrian subduction zone over the last 20 M.y. has been 30 to 40 mm yr^{-1}; the average rate of subduction in the Hellenic subduction zone in the last 10 M.y. has also been 30 to 40 mm yr-1 (However, the actual rate of subduction may have been quite larger in the Hellenic zone over the last 5 M. y.). Thus the Hellenic subduction zone appears to be a mirror image of the Calabrian subduction zone, although the whole process in the east started more than 10 M.y. later than in the west. Consequently, the Aegean sea has not yet reached the oceanization stage and collision has not yet occurred between the southern part of the arc and the opposite Lybian margin.

STRUCTURE AND ORIGIN OF THE EASTERN MEDITERRANEAN BASINS

The structure of the Eastern Mediterranean basins is not as well known as in the west. This is because the basin is filled with very large thicknesses of sediments: 4 to 8 km in the central Ionian basin (Finetti, 1982), 12 km in the central Levantine basin (Makris, 1983). Thus, good quality refraction data were necessary to establish that the deep basins were oceanic. In the central Ionian basin the Moho is 12 to 16 km deep and the crust about 5 km thick (Hinz, 1974; Weigel, 1974). In the central Levantine basin, between Erastosthenes seamount and the Sinaï, the Moho is 25 km deep and the crust 10 to 12 km thick. The crust, there, has a characteristic 6.7 km sec^{-1} velocity which changes to 6.0 km sec^{-1} velocity on reaching the adjacent continental crust (Makris et al., 1983; Ginzburg et al., 1979).

A controversy still exists about the structure of the Mediterranean ridge. This is a large topographic welt which surrounds the Hellenic trench and below which the Moho progressively dips down to a maximum depth of about 30 km equivalent to the depth under the Hellenic arc.

However, it is important to note that this increase in depth of the Moho toward the trench is accompanied by a decrease in free air and Bouguer gravity anomaly. And the decrease is such that it can be explained by assuming that the plate is bent downward, out of isostatic equilibrium, by the subduction process, accompanied by piling up of deformed sediments (Rabinowitz and Ryan, 1970; Le Pichon et al., 1982; Finetti, 1982). Detailed studies of surface topography, using Seabeam (Le pichon et al., 1982) Gloria side scan sonar (Kenyon and Belderson, 1977), deep tow fish (Ryan et al., 1982) confirm that the ridge is affected by extensive shortening, roughly perpendicular to its trend. Rapid uplift of abyssal plain turbidites has been documented (Hsü et al., 1978). Foreward overthrusting over the abyssal plain (Finetti, 1982) and backward thrusting over the Hellenic trench (Huchon et al., 1982; Mascle et al., 1982) have been put into evidence locally.

Thus the ridge is best explained as an accretionary ridge produced above a decollement level which may correspond to an Aptian-Albian shale horizon (Ryan et al., 1982). In this case, the thickness of sediments would increase to a maximum of 15 to 20 km under the Hellenic trench which would then actually be in the position of a fore-arc basin (Le Pichon et al., 1982). The topographic trench would have been buried under the ever expanding accretionary ridge. A comparison with the Barbados ridge has been made (Biju-Duval et al., 1982). Consequently, the determination of a 6 km sec^{-1} velocity over the Mediterranean ridge, mentioned by Makris (1981) but not yet documented in print, may perharps be explained, if it is confirmed, by the presence of limestones, for example. This special structure of the Mediterranean ridge may also be the cause of the continental crustal structure inferred by Cloeting et al. (1980) on the basis of Rayleigh wave dispersion.

At this point, one should note the great similarity in structure between the Calabrian outer ridge and the Mediterranean ridge. In view of the similarity between the Calabrian and Hellenic subduction zones, it is only natural to explain their origin in the same way, by piling up of deformed sediments within an accretionary ridge. The major difference comes from the fact that there is less sediment in the Ionian basin than within the Levantine basin.

To conclude, a concensus has now been reached about the oceanic nature of the deep basins of the Eastern Mediterranean sea. Although there is still a controversy about the exact natures of the Mediterranean ridge and of the Calabrian ridge, both ridges are best explained by an accretionary origin, related to the adjacent subduction. Then, it becomes possible to approximately delineate the extent of the oceanic crust in the Eastern Mediterranean (fig. 2). To the east, one could hesitate to put Erastosthenes seamount within the oceanic crust limit (Ben Avraham et al., 1976; Tanner and Williams, 1984). Many other uncertainties still exist, for example about the exact limit off Egypt and Israël. Figure 2 also shows the position of the limit of the unstretched adjacent continental crust, assuming the same 1.7 stretching ratio as for the Balearic basin margins.

The average heat-flow of the Eastern Mediterranean basins is low and of the order of 30-35 mWm^{-2}, or more probably 35-40 if we take into account the effect of sedimentation (Erickson et al., 1977). This value corresponds to an age significantly larger than 100 My. The isostatically corrected depth in the center of the Ionian basin, based on Finetti (1982), varies between 6.2 and 7.0 km, with 4000 meters of water, 400 meters of Plio-Quaternary, 700 meters of Messinian, 3600 meters of Tertiary and 0 to 2300 meters of Mesozoic. In the Levantine basin, the isostatically corrected water depth is about 6.5 km, based on 2000 meters of water and 11 kilometers of sediments with the densities as given in Makris et al. (1983). This depth range is compatible with ages significantly larger than 100 My. too. Unfortunately, it is not possible, on this basis to argue in favor of either an early Mesozoic or a late Mesozoic age. Thus, the argument about the age must come from other geological considerations although the absence of magnetic lineations (Lort, 1977) suggests a creation during a period without magnetic reversals, that is most probably between Aptian and Santonian (120 to 85 M.y. ago).

Biju-Duval et al. (1977) had pointed out that one could approach this problem by investigating the ages at which the present margins had been submitted to distension and began to form. Biju-Duval and Dercourt (1980) again used this approach to conclude that rifting most probably started in Upper-Triassic-Early Jurassic and that oceanic accretion began in Albian, which is in excellent agreement with the conclusions obtained above. Albian to Senonian is for example the time at which the Syrte basin began to form during a time of major paleogeographic change in Lybia (Burollet et al., 1971). Ryan et al. (1982) have sampled on a mud diapir over the Mediterranean ridge a mud breccia containing benthic deep living forms of Late Aptian age (115 My ago), which suggests that a deep basin existed at this age. Finally, submersible investigations on the margins of the Ionian sea (Escarmed, 1982) indicate that the major paleogeographic change, to the east, on the Malta escarpment, occurred 170 to 180 My ago (Callovo- Oxfordian of Bajocian) when a subsiding platform was rapidly distended and foundered to significant depths. On the African margin of the Cyrenaic plateau, northeast of the Gulf of Syrte, the major paleogeographic change apparently occurred during Middle to Upper Cretaceous, about 100 My ago, as the Senonian is pelagic whereas the Lower to Middle Cretaceous sediments are platform sediments. Upper Jurassic and Upper Cretaceous are known to have been times of distensive basaltic volcanism (Finetti, 1982) and interstratified pillows of these ages have indeed been found on the Malta and the Cyrenaic plateau escarpments.

These submersible investigations support the conclusions made by Biju-Duval and Dercourt (1980). It is known that as oceanic accretion begins, the major extensive tectonic phase ends on the margins because strain gets concentrated along the accretionary boundary. The transition from continental rifting to oceanic accretion is consequently marked by the end of widespread active distension on the margins. The evidence available indicates that the best estimate for the age of this transition around the Eastern Mediterranean is Middle to Upper Cretaceous, after a rather long phase of rifting which began 180 My ago. However, although a younger age can be excluded on the basis of geological and geophysical arguments, one cannot exclude the possibility that oceanic accretion started in Jurassic in aportion of the basin and in Middle-Upper Cretaceous in another part. We had earlier assumed that oceanic accretion, in the Eastern Mediterranean area, was of Jurassic age (Biju-Duval et al., 1977).

THE MESSINIAN SALINITY CRISIS

Although the ages and geologic evolutions of the Mediterranean basins are quite different, as I have shown above, these basins have in common the existence of an evaporitic layer of Messinian age (5 to 7 My ago) which is about 1 kilometer thick as an average (Montadert et al., 1978). It seems that a consensus is now progressively being reached about the mode of deposition of these evaporites in shallow water to completely dissicated deep basins (Hsü et al., 1978). I shall not discuss here the abundant geological and geophysical evidence on this subject (see for a summary Hsü et al., 1978; Montadert et al., 1978; Ryan and Cita, 1978).

Having established that the central parts of the Eastern and Western Mediterranean basins are oceanic and knowing their ages of formation, a reasonable estimate of their depth when the Messinian crisis began 7 My ago can be obtained. It is necessary to unload the Plio-Quaternary and Messinian sediments and to subtract the effect of the thermal subsidence of the lithosphere since Messinian. This last effect is about 300 meters or less in the Balearic basin and about 100 meters or less in the Ionian basin. Using the sedimentary columns given earlier for the Blaearic and Ionian basins, one finds a pre-Messinian paleo-depth of about 3500 meters for the Balearic basin and 4500 meters for the Ionian basin. However, one should make a correction for the decompaction of sediments which may decrease these depths to about 3200 and 4300 meters respectively. In the absence of water during Messinian dessication periods, the depth would have been 2200 meters and 2900 meters respectively. Unless one assumes that the law of thermal subsidence of the oceanic lithosphere in the Mediterranean basins is completely different from other marginal basins in the world, one cannot escape the conclusion that the water depth 7 My ago, before the salinity crisis, was not smaller but rather larger than today, both in the Blearic and the Ionian basin. This was the conclusion reached on the basis of similar arguments by Le Pichon et al. (1971), Hsü et al. (1973) and Ryan (1976). It is especially difficult to escape this conclusion in the center of the Ionian basin, which is clearly made of old oceanic lithosphere in isostatic equilibrium and where present water depth is large (about 4000 meters) and combined Plio-Quaternary and Messinian thickness is 1 to 1.5 km.

The widespread evidence for erosional surfaces on the margins of the basins and for deeply incised canyons below the largest rivers (Ryan and Cita, 1978; Hsü et al., 1978) is compatible with this conclusion.

I now wish to briefly discuss the tectonic effect of unloading and loading about 3.5 km of water upon the lithosphere of these basins. An unloading of the same amplitude has occurred at the end of the glacial stages. However, in the case of the Mediterranean, the limits of the load coincide with a narrow geological boundary, the continental margin, and do not migrate with time. This difference is crucial because the stresses related to flexuration on the edges of the load are concentrated and have tectonic consequences. It is known that the deformation of the lithosphere under loads can be approximated using a thin elastic plate model. For the oceanic lithosphere, the thickness h of the equivalent oceanic thin plate varies approximately as $h = 3A^{1/2}$ km, where A is the age in millions of years (Watts et al., 1982). Thus, when the Mediterranean lithosphere was rapidly unloaded of 35oo meters of water, the seafloor moved upward to a new level corresponding to a depth of 2400 meters whereas the adjacent continet did not move. This differential motion of about one kilometer was absorbed by elastic flexure over the widths of the continental margins (Watts and Ryan, 1976). Le Pichon et al. (1971) proposed that such a rapid flexure, localized along the continental margins, must have resulted in important stresses which might have induced tectonics movement.

Let us assume that the Mediterranean lithosphere obeys the general relationship between h and A proposed by Watts et al. (1982). Then, the equivalent thickness of the elastic plate is about 13 kilometers for the Balearic basin and 35 kilometers for the Ionian basin. The corresponding flexural parameter is respectively 40 and 80 km. It is eay to compute the deflexions and the stresses induced by the deflexions for a thin elastic plate having these values for the flexual parameter. This has been done in figure 4 for a basin 3500 meters deep, with a continental margin about 50 kilometers wide as in the Gulf of Lions, assuming two-dimensionality.

The resulting topographic uplift of the edge of the continental shelf during unloading was previously noted by Le Pichon et al.(1971) and Watts and Ryan (1976). Note that, as the deflexions are added on top of an existing very large topography, the locations of the maximum uplift are always exactly on the edge of the shelf, independently of the value of the flexural parameter. Note on the other hand that the locations of the maximum stresses do not necessarily coincide with the shelf edge and the base of the continental slope. This coincidence occurs for the West Mediterranean basin when the continental margin is 50 kilometers wide and for the East Mediterranean basins when it is 100 km wide which are values quite typical of young and mature continental margins respectively. In this case, horizontal compressive stresses of about 100 MPa (1000 bars) are concentrated at the basis of the continental margins whereas equivalent horizontal extensional stresses occur on the shelf edge. These are large stresses They are localized in such a way that they might lead to overthrusting of the margin over the adjacent basin if an additional regional compressive tectonic stress field does exist.

Do we have any evidence for such overthrusting during Messinian? I suggest that the greatly increased Hellenic subduction at this time is such an evidence (Laj et al., 1982). Other evidence for compressional tectonics at the basis of the continental margin and for simultaneous extensional tectonics on the continental slope may actually exist all along the northern Eastern Mediterranean margins.

CONCLUSIONS

The present Mediterranean seas consist in part of a relic probably Middle Cretaceous oceanic basin which began to be subducted from NW to SE under what is now the Calabrian arc about 25 to 30 My ago and from NE to SW under the Hellenic arc about 10 My ago. Subduction is still active today. It has resulted in the formation of the oceanic Lower Miocene Balearic and North Algerian basins and the semi-oceanic Upper Miocene to Plio-Pleistocene Tyrrhenian basin behind the Calabrian arc, and in the formation of the continental Aegean sea behind the Hellenic arc. The Aegean sea, today, is at the stage at which the Tyrrhenian Sea was during Upper Miocene.

The evolution of the whole area is dominated by the dynamics of subduction. As continental collision between Europe and Africa proceeded at its slow pace of less than 1 cm yr^{-1}, overthrusting of the continental margin over the adjacent Middle Cretaceous oceanic basin was initiated first to the extreme-west, somewhere near the Alboran plate, then to the extreme-east, between Crete and Rhodes. Subduction was able to start and reach a steady velocity of 3-4 cm yr^{-1}, whereas continental collision continued at its much slower pace. This fact is an indication that the "trench pull" force dominates the tectonic evolution although the push forces due to lateral density variations within the stretching basins may also play a significant role (Le Pichon,1982). The rapid subduction then necessarily leads to adjacent continental stretching and subsidence and consequent formation of a new basin in the wake of the disappearing older one until collision occurs between the margins of the older basin. Because of a complex geometry, collison may occur in one part of the basin, while subduction and extension continue elsewhere. However, a tectonic cycle finally ends when the older basin has been completely subducted.

The amount of extension in the stretched basin is related to the amount of subduction in the adjacent subduction zone. Using this argument, I have shown that the present length of the Calabrian subduction zone indicates that there has been large amounts of continental crust stretching in the Western Mediterranean basins, amounts comparable to the ones computed by Le Douaran et al. (1984) assuming conservation of continental crust. This is an argument against large scale subsidence without stretching by continental crust assimilation in the mantle.

Finally, the Messinian salinity crisis occurred in basins which were deeper than today and the unloading of water induced significant tectonic stresses on the margins. These stresses might have led to overthrusting of the margins over the adjacent basins.

ACKNOWLEDGEMENTS

G.Boillot, M.Recq, L.Montadert and C.A. Williams communicated preprints. Discussions with J.Dercourt, C.Grandjacquet and L.E. Ricou were quite useful. N.Chamot-Rooke made the computations of Figure 4. Research was financed by CNEXO and CNRS.

REFERENCES

AGIP, Carte magnetica d'Italia, anomalie del campo magnetico residuo; 1.500.000, 1981.

Allegre C.J., Courtillot V. et al., Structure and evolution of the Himalaya-Tibet orogenic belt, Nature, 307, 17-22, 1984.

Andrieux J., Fontbote J.M. and Mattauer M., Sur un modèle explicatif de l'arc de Gibraltar, Earth Planet.Sci.Lett., 12, 191-198, 1971.

Avedik F., Camus A.L., Ginsburg A., Montadert L., Roberts D.G. and Whitmarsh R.B., A seismic refraction and reflexion study of the continent-ocean transition beneath the North Biscay margin, Phil.Trans. R.Soc. London, A 305, 5-25, 1982.

Banda E., Ansorge J., Boloix M. and Cordoba D., Structure of the crust and upper mantle beneath the Balearic islands (Western Mediterranean), Earth Planet.Sci. Lett., 49, 219-230, 1980.

Barberi F., Bizouard H., Capaldi G., Ferrara G., Gasparini P., Innocenti F., Joron J.L., Lambret B., Treuil M. and Allegre C., Age and nature of basalts from the Tyrrhenian abyssal plain, Initial Reports DSDP, 42 A, Washington, 509-514, 1978.
Bayer R., Le Mouel J.L. and Le Pichon X., Magnetic anomaly pattern in the Western Mediterranean, Earth Planet. Sci.Lett., 19, 168-176, 1973.
Ben Avraham Z., Shoham Y. and Ginsburg A., Magnetic anomalies in the Eastern Mediterranean and the tectonic setting of Erastosthenes seamount, Geophys.J.R.Astr.Soc., 45, 105-123, 1976.
Biju-Duval B. and Dercourt J., Les bassins de la Méditerranée orientale représentent-ils les restes d'un domaine océanique, la Mésogée, ouverture Mésozoïque et distincte de la Téthys? Bull.Soc.Géol.France, 7, 43-60, 1980.
Biju-Duval B. and Montadert L., ed., Structural history of the Mediterranean seas, Technip, 445 p., 1977.
Biju-Duval B., Dercourt J. and Le Pichon, X., From the Tethys ocean to the Mediterranean seas. In "Structural history of the Mediterranean basins", Technip, Rueil Malmaison, 143-164, 1977.
Biju-Duval B., Le Quellec P, Mascle A., Renard V., and Valery P., Multibeam bathymetric survey and high resolution seismic investigations on the Barbados Ridge complex, Tectonophys., 86, 275-304, 1982.
Bousquet J.C. and Philip H., Les caractéristiques de la néotectonique en Méditerranée occidentale, in F.C.Wezel, ed., Sedimentary basins of Mediterranean margins, Tecnoprint, Bologna, 389-405, 1981.
Burollet F., Magnier P., and Manderscheid C., La Lybie, in Tectonique de l'Afrique, Unesco, Paris, 409-416, 1971.
Burrus J., Contribution to a geodynamical synthesis of the Provencal Basin, Tectonophysics, in press, 1984.
Cloeting S., Nolet G. and Wortel R., Crustal structure of the Eastern Mediterranean inferred from Rayleigh wave dispersion, Earth Planet.Sci.Lett., 51, 336-342, 1980.
Durand-Delga M., Ouvertures océaniques de Méditerranée occidentale et dislocation des chaînes alpines, in F.C. Wezel, ed., Sedimentary basins of Mediterranean margins, Tecnoprint, Bologna, 417-432, 1981.
Edel J.B., Montigny R. and Thuizat R. Late Paleozoic ro-

tations of Corsica and Sardinia: new evidence from paleomagnetic and K/Ar studies, Tectonophys., 79, 201-223, 1981.

Erickson A.J., Simmons G. and Ryan W.B.F., Review of heat-flow data from the Mediterranean and Aegean seas, In "Structural History of the Mediterranean basins", B.Biju-Duval and L.Montadert ed., Technip, Rueil-Malmaison, 263-279, 1977.

Escarmed, Données nouvelles sur les marges du Bassin ionien profond (Méditerranée orientale), résultats des campagnes Escarmed, Rev.Inst. Francais Pétrole, 37, 713-731, 1982.

Estocade, Messinian subaerial erosion of the Stoechades and Saint-Tropez canyons, a submersible study, Marine Geology, 27, 247-269, 1978.

Fabbri A. and Curzi P., The Messianian of the Tyrrhenian sea: seismic evidence and dynamic implications, Giornale di Geologia, 2, 215-248, 1979.

Finetti I., Structure, stratigraphy and evolution of Central Mediterranean, Bolletine di Geofisica Teorica ed Applicata, 29, 247-315, 1982.

Finetti I. and Morelli C., Geophysical exploration of the Mediterranean Sea, Boll.Geof. Teor., Appl., 15, 263-344, 1973.

Galdeano A., Courtillot V., Le Borgne E., Le Mouel J.L., Rossignol J.C., Aeromagnetic survey of the Mediterranean description and tectonic implications, Earth Planet.Sci. Lett., 23, 323-336, 1974.

Gasparini C., Iannacone I., Scandone P. and Scarpa R. Seismotectonics of the Calabrian arc, Tectonophysics, 84, 267-286, 1982.

Ginsburg A., Makris J., Fuchs F., Prodehl C., Kaminski W. and Amitai V., A seismic study of the crust and upper mantle of the Jordan-Dead Sea Rift and their transition toward the Mediterranean Sea, J.Geophys.Res., 84, 1569-1582, 1979.

Hatzfeld D., Etude sismotectonique de la zone de collision ibéro-maghrébine, These Doc.Sci., Univ. Grenoble, 281 p., 1978.

Hinz K., Crustal structure of the Balearic Sea, Tectonophys., 20, 295-302, 1973.

Horvath F., Berckhemer H. and Stegena L. Models of Mediterranean back-arc basin formation, Phil.Trans.R.Soc. London, A300, 383-402, 1982.

Hsü K.J., Tectonic evolution of the Mediterranean basins, in "The oceans basins and margins" 4A, Ed. A.E.M.Nairn, W.H.Kanes, F.G.Stehli, Plenum, New York, 29-75, 1978.

Hsü K.J., Cita M.B. and Ryan W.B.F., The origin of the Mediterranean evaporites, Deep Sea Drilling Rep. 13, 1203-1231, 1973.

Hsü K.J., Montadert L. et al., Initial reports of the Deep Sea Drilling Project, 42, I, Washington, D.C., 1978.

Hsü K.J., Montadert L., Bernoulli D., Cita M.B., Erickson A., Garrison R.E., Kidd R.B., Melieres F., Muller C., and Wright R., History of the Mediterranean salinity crisis, In: Init.Reports of DSDP, 42, Washington D.C., 1053-1079, 1978.

Hutchinson I., Heat-flow studies in the Gulf of Oman and in western Mediterranean basins, PhD Thesis, Cambridge, 1983.

Kenyon V.M. and Belderson R.H.B., Young compressional structures of the Calabrian, Hellenic and Cyprus outer ridge, In Structural History of the Mediterranean basins, Technip, Rueil, 233-240, 1977.

Kissel C., Jamet M. and Laj, C. Paleomagnetism evidence of Miocene and Pliocene rotational deformations of the Aegean area, J.Geol.Soc. London, in press, 1984.

La Brecque J.L., Kent V., Cande S.C., Revised magnetic polarity time scale for late Cretaceous and Cenozoic time, Geology, 5, 330-335, 1977.

Laj C., Jamet M., Sorel D. and Valente J.P., First paleomagnetic results from Mio-Pliocene series of the Hellenic sedimentary arc, Tectonophysics, 86, 45-67, 1982.

Leblanc D.and Olivier P., Role of strike-slip faults in the Betic-Rifan orogeny, Tectonophys. 101, 345-355, 1984.

Le Douaran S., Burrus J. and Avedik F., Deep structure of the North Western Mediterranean basin : results of a two ships seismic survey, Tectonophysics, in press, 1984.

Letouzey J. and Tremolieres P., Paleo-stress fields around the Mediterranean since the Mesozoic derived from microtectonics : comparisons with plate tectonic data, in 26th Int. Geol. Cong., Colloquium N°5, BRGM ed., 261-273,1980.

Le Pichon, X. Land-locked oceanic basins and continental collision: the Eastern Mediterranean as a case example, in Mountain Building Processes, K.J.Hsü ed., Academic Press, 201-212, 1982.

Le Pichon X. and Angelier J. The Hellenic arc and trench system: a key to the neotectonic evolution of the Eastern Mediterranean area, Tectonophysics, 60, 1-42, 1979.

Le Pichon X. and Anglelier J. The Aegean Sea, Phil.Trans. Roy.Soc.London, A.300, 357-372, 1981.

Le Pichon X. and Sibuet J.C., Passive margins: a model of formation, J.Geophys.Res., 86, 3708-3720, 1981.

Le Pichon X. Augustithis S.S. and Mascle J., ed., Geodynamics of the Hellenic arc and trench, Tectonophysics, 86, 1982.

Le Pichon X., Lyberis N. and Alvarez F., The North Aegean trough: subsidence, J.Geol.Soc.London, in press, 1984.

Le Pichon X., Pautot G. and Weill J.P., Opening of the Alboran Sea, Nature, 236, 83-85, 1972.

Le Pichon X., Lyberis N., Angelier J. and Renard V., Strain distribution over the East Mediterranean ridge: a synthesis incorporating new Sea-beam data, Tectonophys. 86, 243-274, 1982.

Le Pichon X., Pautot G., Auzende J.M., and Olivet J.L., La Méditerranée occidentale depuis l'Oligocène, schéma d'évolution, Earth Planet. Sci.Lett., 13, 145-152, 1971.

Lort J.M., Geophysics of the Mediterranean Sea basins, In A.E.M.Nairn, W.H.Kanes, F.G.Stehli, The ocean basins and margins, 1A, The Eastern Mediterranean, 151-213, Plenum, New York, 1977.

Louden K.E., The crustal and lithospheric thicknesses of the Philippine Sea as compared to the Pacific, Earth Planet.Sci.Lett. 50, 275-288, 1980.

Makris J., Deep structure of the Eastern Mediterranean deduced from refraction seismic data, in Sedimentary Basins of Mediterranean margins, F.C.Wezel ed., Tecnoprint, Bologna, 63-64, 1981.

Makris J., Ben Avraham Z., Behle A., Ginsburg A., Giese P. Steinmetz L., Whitmarsh R.B., and Eleftheriou S., Seismic refraction profiles between Cyprus and Israël and their interpretation, Geophys. J.R.Astr.Soc., 75, 575-591, 1983.

Mascle J., Jongsma D., Campredon R., Dercourt J., Glacon

G., Lecleach A., Lyberis N., Malod J.A. and Mitropoulos, D., The Hellenic margin from Eastern Crete to Rhodes: preliminary results, Tectonophys., 86, 133-147, 1982.

McKenzie D.P. Active Tectonics of the Alpine-Himalayan belt: the Aegean sea and surrounding regions (Tectonics of Aegean region), Geophys.J.R.Astron.Soc., 55, 217-254, 1978.

Meissner R. and Strehlau J., Limits of stresses in continental crusts and their relation to the depth-frequency distribution of shallow earthquakes, Tectonics, 1, 73-89, 1982.

Molnar P., Freedman D. and Shih J.S. Lengths of intermediate and deep seismic zones and temperatures in downgoing slabs of lithosphere, Geophys. J.R.Astr.Soc., 56, 41-54,1979.

Montadert L., Letouzey J. and Mauffret A. Messinian event: seismic evidence, Initial Reports of the Deep Sea Drilling Project, 42 A, Washington, 1937-1050, 1978.

Moussat E. Evolution de la mer tyrrhénienne centrale et orientale et de ses marges en realtion avec la néotectonique dans l'arc calabrais, Thèse de 3e cycle, Université Pierre et Marie Curie, Paris, 122 p., 1983.

Nairn A.E.M., Kanes W.H. and Stehli F.G. ed., The ocean basins and margins, Volumes 4A and 4B, Plenum, New York, 1977.

Olivet J.L., Bonnin J., Beuzart P.and Auzende J.M. Cinématique de l'Atlantique Nord et Central, Notice, in Rapports Scientifiques CNEXO, Paris, in press, 1984.

Papazachos B.C. Distribution of seismic foci in the Mediterranean and surrounding areas and its tectonic implications, Geophys.J.R.Astron.Soc., 33, 421-430, 1973.

Papazachos B.C. Evidence of crustal shortening in the Northern Aegean region, Boll.Geofis. Teor.Appl., 19, 66, 1976.

Rabinowitz D. and Ryan W.B.F. Gravity anomalies and crustal shortening in the Eastern Mediterranean, Tectonophys. 10, 585-608, 1970.

Recq M., Rehault J.P., Steinmetz L. and Fabbri A., Amincissement de la croûte et accrétion au centre du bassin tyrrhénien d'après la sismique réfraction, Marine Geology, in press, 1984.

Rhehault J.P., Boillot G. and Mauffret A. The Western

Mediterranean basins, Tectonophysics, in press, 1984.
Ryan W.B.F., Quantitative evaluation of the depth of the Western Mediterranean before, during and after the late Miocene salinity crisis, Sedimentology, 23, 791-813, 1976.
Ryan W.B.F. and Cita M.B. The nature and distribution of Messinian erosional surfaces-indicators of a several-kilometer-deep Mediterranean in the Miocene, Marine Geology, 27, 193-230, 1978.
Ryan W.B.F., Katens K.A. and Cita M.B. Geological evidence concerning compressional tectonics in the Eastern Mediterranean, Tectonophys., 86, 213-242, 1982.
Selli R. Profili magnetometrici, in : R.Selli, ed., Ricerche Geologiche Preliminari nel Mar Tirreno, Giorn. Geol., 37, 43-54, 1970.
Steinmetz L., Ferucci F., Hirn A, Morelli C and Nicolich R, A 550 km long Moho traverse in the Tyrrhenian Sea from O.B.S. recorded PN waves, Geophys. Res.Lett., 10, 428-431, 1983.
Tanner S.J. and Williams C.A. A detailed survey near mount Erastosthenes, Eastern Mediterranean, Marine Geophys. Res. in press, 1989.
Tapponnier P. Evolution tectonique du système alpin en Méditerranée : poinconnement et écrasement rigide-plastique. Bull.Soc. Geol.Fr., 19, 437-460, 1977.
Watts A.B. and Ryan W.B.F. Flexure of the lithosphere and continental margin basins, Tectonophys., 36, 25-44, 1976.
Weigel W. Crustal structure under the Ionian Sea, J.Geophys., 40, 137-140, 1974.
Westphal M., Bazhenov M.L., Lauer J.P., Pechersky D.M. and Sibuet J.C., Paleomagnetic implications on the evolution of the Tethys belt from the Atlantic ocean to Pamir since Trias, Tectonophysics, submitted, 1984.
Wezel F.C. ed., Sedimentary basins of Mediterranean margins, Tecnoprint, Bologna, 520 p., 1981.

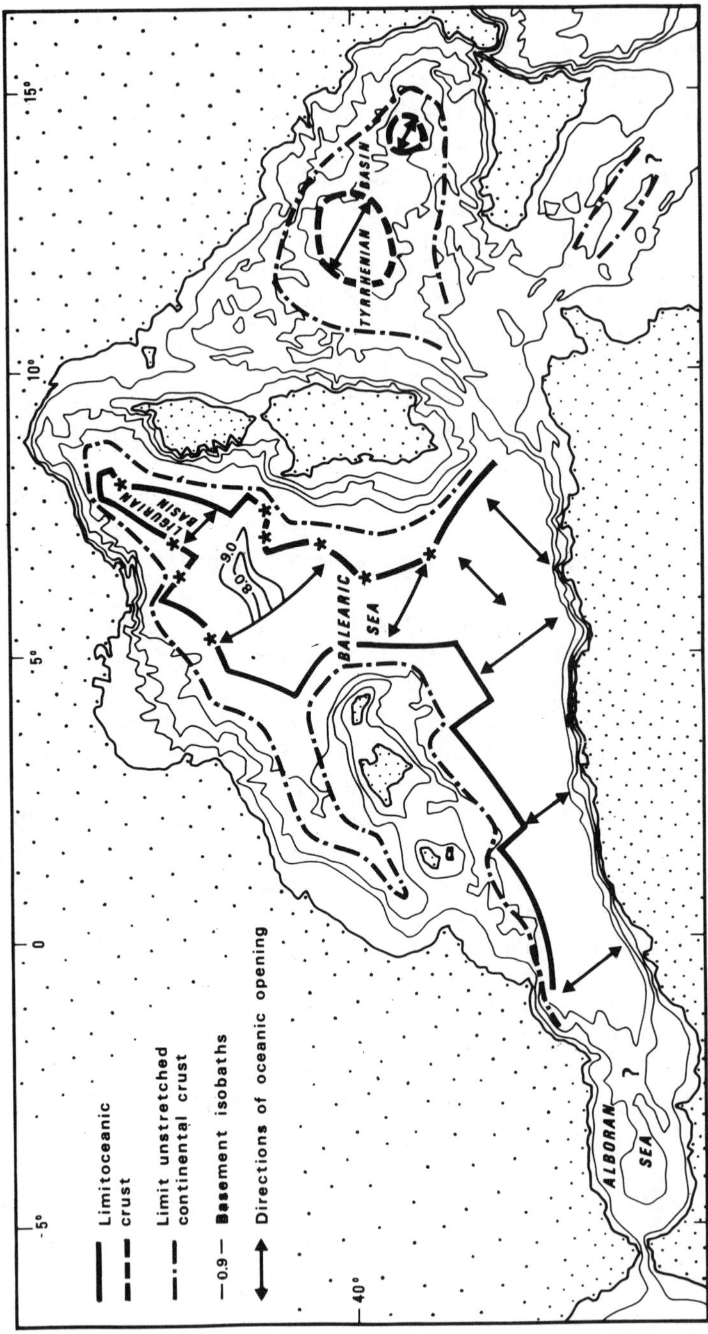
Limitoceanic crust
Limit unstretched continental crust
—0.9— Basement isobaths
Directions of oceanic opening
LIGURIAN BASIN
TYRRHENIAN BASIN
BALEARIC SEA
ALBORAN SEA
9.0
8.0
-5°
0
5°
10°
15°
40°

FIGURE 1. Oceanic-continental crust boundary in the Western Mediterranean basins. The limit in the Balearic Sea is after Burrus (1984) and is well defined. It has been extrapolated to the south in the North Algerian basin where its exact location is still hypothetical. In the Tyrrhenian basin, the limit outlines areas of extremely thin crust (5 km or less). The limit of unstretched continental crust has been obtained by assuming continental crust conservation and a ratio of 1.7 between stretched and unstretched crust, from the shelf edge to the oceanic drust limit. Basement isobaths outline the crest of the fossil spreading zone in the Balearic basin. The directions of oceanic opening are mostly based on magnetic anomalies interpretation.

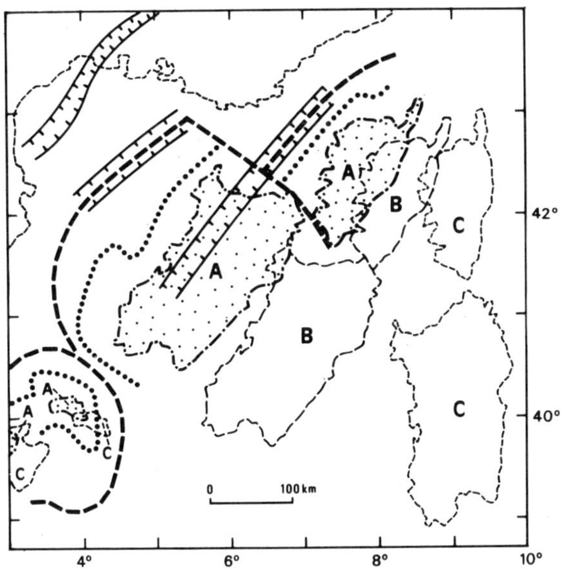

FIGURE 2. Details of Corsica-Sardinia rotation. "C" corresponds to present and "B" to the position at the beginning of oceanic accretion, after Burrus (1984). The position of "A" is obtained by unstretching the continental margin using the quantative estimates of Le Douaran et al. (1984). This result is locally different from the one proposed in figure 1, where it was assumed that stretching had not occurred beyond the edge of the continental shelf. Late Oligocene major lines of Rifting are indicated.

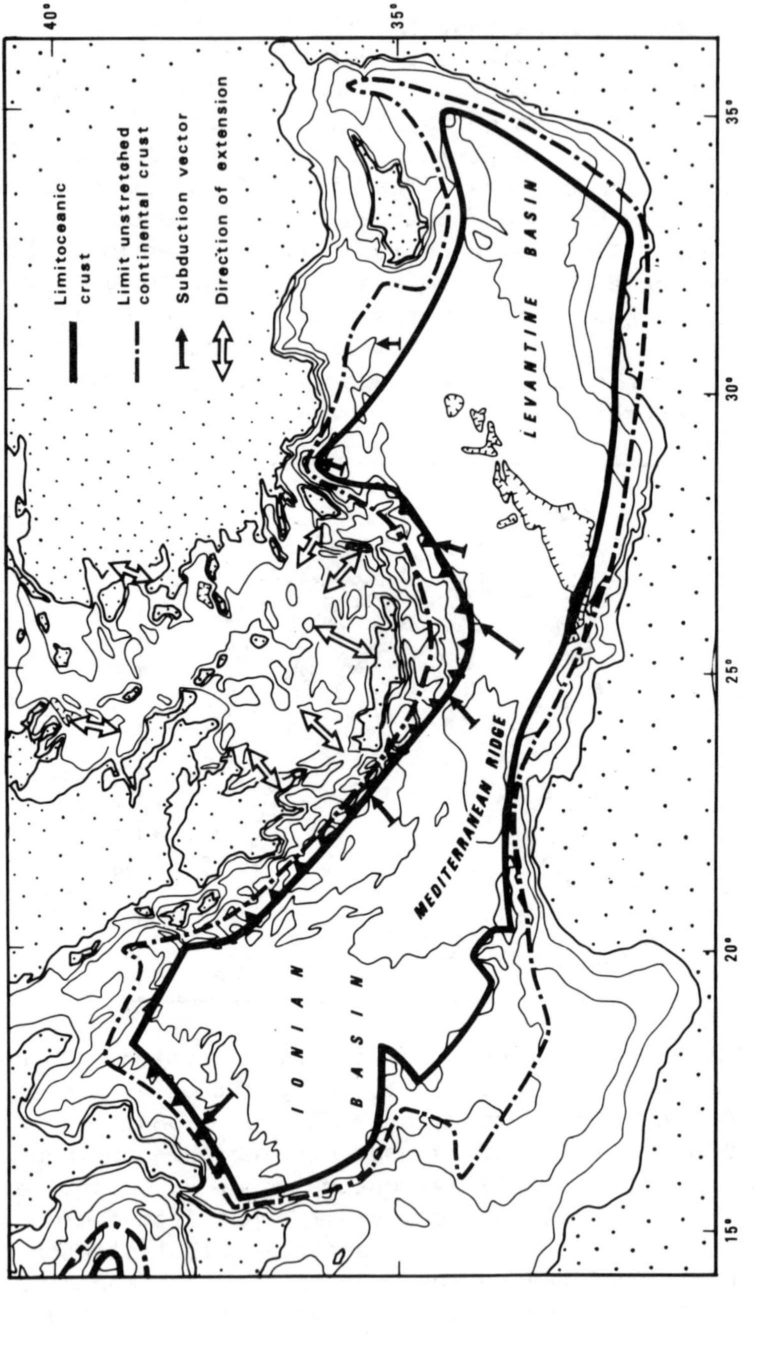
Limitoceanic crust
Limit unstretahed continental crust
Subduction vector
Direction of extension
IONIAN BASIN
MEDITERRANEAN RIDGE
LEVANTINE BASIN
40°
35°
15°
20°
25°
30°
35°

FIGURE 3. Oceanic-continental crust boundary in the Eastern Mediterranean basins. This boundary is based on crustal thickness, extrapolated from seismic refraction and gravity measurements in the central IOnian and Levantine basins and its exact location is still hypothetical. It is further assumed that the lithosphere now subducted under the Calabrian and Hellenic arcs is oceanic for reasons discussed in the text. The limit of unstretched continental crust was obtained in the same way as for figure 1. The subduction vectors are after Le Pichon and Angelier (1981) for the Hellenic arc. The directions of extension in the Aegean area are also after Le Pichon and Angelier (1981).

FIGURE 4. Elastic flexure along the continental margins of the Mediterranean basins during Messinian dessication assuming a 3500 meters pre-Messinian water depth. Continuous curve for the Balearic basin, dashed curve for the Ionian basin. Positive stresses are extensional and negative stresses are compressional.

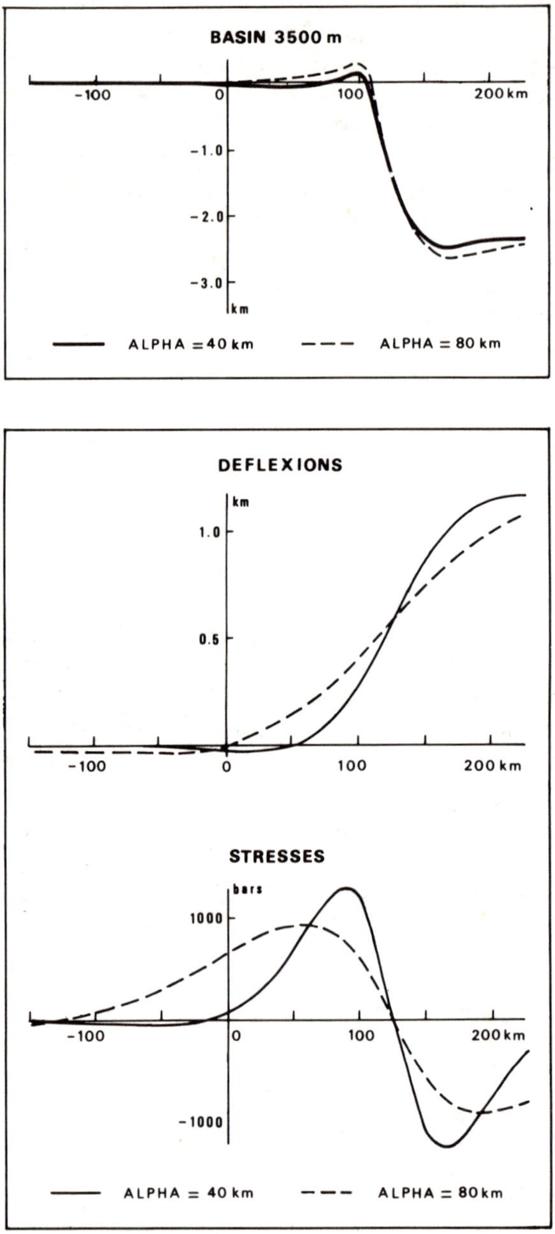

FIGURE 4. Caption on previous page.

specific gravity and other physical-chemical properties, these fluids differently reacting with rocks; 3) because one and the same layer occurring in different structures and/or in different parts of one and the same structure is under different P-T conditions; 4) due to irregularities in distribution of the tectonic stresses in the Earth Crust as well as within the limite of separate structures (1,10,12,21,22,26).

One should add the compactive unevenness of different rock types with depth which makes a certain contribution to formation of structures (18). Flexure of bed sequences of variable thickness leads to formation of rootless structures and structures with displaced crests (Fig. 2a, b) (24), which sometimes are interpreted as sedimental ones (17). Composition of sediments plays an important role in the nature of dislocation. Thus, in tectonically active regions of the Middle Asia very narrow ("sharp") folds are not observed in hydromicaceous clays but widely developed in sequences, containing layers of montmorillonite clays. At present all the groups of processes mentioned above are thoroughly studied but very seldom one tries to attribute them to concrete structural forms of lithosphere or study tectonic control of their manifestation.

So, between the structural tectonic factor and the lithogenesis at all its stages there exist very close, complicated and often inverse relations which are not properly taken into consideration when making concrete lithogenetic investigations.

Further investigation should be devoted to typifying lateral variability of sedimentary and catagenetic features of rocks and their physical properties on tectonic structures and revealing the dynamics of interrelation between tectogenesis and lithogenesis in detail, taking concrete regional and local structures as examples.